MINERALS, ENERGY, AND ECONOMIC DEVELOPMENT IN CHINA

MINERALS, ENERGY, AND ECONOMIC DEVELOPMENT IN CHINA

JAMES P. DORIAN

CLARENDON PRESS · OXFORD
1994

Oxford University Press, Walton Street, Oxford OX2 6DP
Oxford New York Toronto
Delhi Bombay Calcutta Madras Karachi
Kuala Lumpur Singapore Hong Kong Tokyo
Nairobi Dar es Salaam Cape Town
Melbourne Auckland Madrid
and associated companies in
Berlin Ibadan

Oxford is a trade mark of Oxford University Press

Published in the United States
by Oxford University Press Inc., New York

British Library Cataloguing in Publication Data
Data available

Library of Congress Cataloging in Publication Data
Dorian, James P.
Minerals, energy, and economic development in China / James P. Dorian.
Includes bibliographical references and index.
1. Mineral industries—China. 2. Mines and mineral resources—China. I. Title.
HD9506.C62D67 1994 93–38038
338.2'0951—dc20
ISBN 0–19–828744–5

1 3 5 7 9 10 8 6 4 2

Typeset by Best-set Typesetter Ltd., Hong Kong
Printed in Great Britain
on acid-free paper by
Biddles Ltd., Guildford and King's Lynn

To
My Family and Friends

PREFACE

LIKE the four rounds of international bidding for China's offshore oil, much of the nation's mining development history has been subject to erratic behaviour and uncertainty. Even today, despite the realization that China possesses one of the largest mining industries in the world, the future course of minerals and energy exploration and development in the country is difficult to assess, particularly in terms of foreign participation in such activities. To forecast long-term mining achievements in China reliably, a thorough understanding of historical developments is required.

Since China opened its doors to the world in the late 1970s, the nation has continued to grow in importance in the global economic equation. Its economic reforms are being watched with great interest all over the world, and particularly by other nations of Asia because of their proximity to this most populous country on earth. Civil unrest in 1989 caused great concern among the international community, but foreign investment continues to pour into China at record levels.

This book attempts to describe, in a consolidated manner, the major characteristics and qualities of the Chinese mining industry based on its past and present performance. China's rich endowment in mineral and energy resources and large untapped potential labour force and consumer markets combine to make the nation's mining activities and policy directions highly significant to the world's minerals and energy balance. The various options open to China's mining industry for future development activities are also analysed in the book with regard to the many constraints on the industry today. The Chinese mining industry is extremely large and complex; this book answers some of the questions about it, but raises many as well. Only the future direction of the industry will bring answers to these questions.

It is nearly impossible to write a book on China's mining industry without the assistance and support of numerous individuals and groups. The eight chapters in this book are based, in large part, on my many travels to the People's Republic since 1985. During these visits I had the fortunate opportunity of meeting with many senior Chinese government and mining industry officials. I wish to thank the many groups and organizations that answered so many of my questions during these trips. They include the China National Nonferrous Metals Industry Corporation; the Ministry of Geology and Mineral Resources; the Ministry of Metallurgical Industry; the Ministry of Energy; Wuhan College of Geology; Beijing University; the State Science and Technology Commission; the State Planning Commission; the Institute of Geology, Chinese Academy of Sciences; the State Statistical Bureau; Xinjiang Engineering Institute; Xinjiang Bureau of Geology and Mineral Resources; the Research Centre for Technological, Economic, and Social

Development of the State Council; and the State Price Bureau. I would like to personally thank Liu Youxi, formerly with CNNC, for his numerous candid discussions with me about the minerals base and industry of China.

My place of employment, the East–West Center, is gratefully acknowledged for providing me the opportunity and financial support to write this book. At that institution, I would especially like to thank my closest research colleagues, Dr Allen L. Clark, Dr Charles J. Johnson, and Dr James M. Otto from the Program on Resources: Energy and Minerals for their continued guidance and support throughout this project. With them, I travelled throughout China visiting fascinating mining areas and geologic sites. Dr Seiji Naya, the former Director of the Center's Resource Systems Institute, is recognized for both supporting my research and critiquing an early draft of this book. Ye Feng and Zhang Shuqiang provided invaluable help to me in Chinese translation services, and Patricia Wilson, former East–West Center Program Representative in Beijing, spent countless hours scanning Chinese publications for mineral-related articles. Dr James Nickum, formerly with Winrock International's China Office and now with the East–West Center's Program on Environment, is thanked for his helpful advice on the linkages between agriculture, rural industry, and mining in China. Lillian Shimoda, Dee Baroman, and Dave Miller of the Center are gratefully acknowledged for their help in preparation of the manuscript, and Laurel-Lynn Indalecio is proudly recognized for her graphics drawing. D. Lyn Henagan and Wilfredo Icay are acknowledged for their unparalleled computer programming. David Fridley of the East–West Center's Program on Resources: Energy and Minerals is commended for the extensive and invaluable help he gave to me with this book. Indeed, writing the book would not have been possible without the guidance and support of Mr Fridley.

Other persons and organizations who in some way assisted in the preparation of this book include Dr Jeffrey Taylor of the US Department of Commerce, who thoroughly critiqued the input–output analysis presented in Chapter 3, Edmond Chin of the US Bureau of Mines, who provided some of the background data used in the book, and the US–China Business Council in Washington, DC, who opened their files to me as a doctoral candidate so many times. An anonymous reviewer for Oxford University Press is gratefully acknowledged for constructive comments and criticisms on preliminary drafts of this book. None of the above should be held responsible in any way for errors or misguided opinions in what follows.

J.P.D.

Honolulu, Hawaii

CONTENTS

FIGURES

TABLES

ABBREVIATIONS

ASEAN	Association of Southeast Asian Nations
b/d	barrels per day
CCP	Chinese Communist Party
CGE	computable general equilibrium
CIISIC	China International Iron and Steel Investment Corporation
CMIEC	China Metallurgical Import–Export Corporation
CNNC	China National Nonferrous Metals Industry Corporation
CNOOC	China National Offshore Oil Company
CNPC	China National Petroleum Corporation
DFI	direct foreign investment
GATT	General Agreement on Tariffs and Trade
GRIMP	GRit IMpact Program
GRIT	Generation of Regional Input–Output Tables
GVAIO	gross value of agricultural and industrial output
GVAO	gross value of agricultural output
GVIO	gross value of industrial output
MBS	Material Balancing System
MINMETALS	China National Metals and Minerals Import and Export Corporation
mmb/d	million barrels per day
MOFERT	Ministry of Foreign Economic Relations and Trade
mtce	million tonnes of coal equivalent
mtoe	million tonnes of oil equivalent
mt/y	million tonnes per year
NIE	newly industrializing economy
PLA	People's Liberation Army
PRC	People's Republic of China
RMB	renminbi (yuan)
SINOCHEM	China National Chemical Import and Export Corporation
SINOPEC	China National Petrochemical Corporation
SITC	Standard International Trade Classification
SNA	System of National Accounts
TCCR	Tentative Criteria for Mineral Reserves Classification
Twh	terawatt-hours
t/y	tonnes per year

PART I

MINING AND DEVELOPMENT PLANNING

1
Introduction

MINERALS AND ECONOMIC DEVELOPMENT IN CHINA

Today, China possesses one of the world's largest mining industries,[1] ranking first in world coal production,[2] sixth in petroleum output, and is among the leading producers of several metallic and non-metallic resources. The Chinese mining industry is extraordinarily complex, having evolved into an overwhelmingly large bureaucracy which has been, until recently, both shrouded from and misunderstood by the international community. Although to date more than 160 types of minerals have been discovered in China, a significant portion of the minerals and energy base remains largely unidentified and evaluated. Despite China's large reserves of several mineral resources, with a population of more than 1.1 billion the country ranks among the lowest in the world in terms of proved reserves per capita. In addition, mining recovery rates in China average about 50 per cent, significantly lower than the world average of near 90 per cent.

As providers of raw materials and substantial foreign exchange earnings, China's energy, metal, and non-metal industries are crucial to the country's economic development. These industries have been undergoing enormous changes since the economic reform programmes began in 1978. The pre-1978 period was characterized by a policy of self-reliance for minerals and energy, uncoordinated and imbalanced development of these industries, and limited external trade. In contrast, after 1978 China looked to these industries to be driving forces for the country's entry into the international economy. It actively sought foreign capital and entered into joint ventures to modernize and expand mineral and energy production and exports.

Despite successes in many areas, Chinese leaders have found the development of China's mineral and energy production and exports to be an especially difficult challenge. In the past decade, three important changes in the country's mineral and energy outlook have affected Chinese policy. First, the Chinese have realized that, despite China's rich resource endowments, growing domestic demand will absorb an increasing share of production and is causing China to become a major importer of some mineral products. Secondly, depressed demand and prices for many resources in world markets have made mineral and energy exports less attractive as earners of foreign exchange. Thirdly, China's ability to expand supplies encountered serious difficulties. As an example, the exploration and development of offshore oil has been disappointing. Other constraints in the development and utilization of mineral and energy commodities have included transportation bottlenecks, the inability to modernize existing facilities rapidly enough to expand production, and lower-than-expected joint-venture investment in the mineral and energy sectors.

Future policies and strategies are needed that will encourage further expansion of China's mining industry. China's well-publicized goal of quadrupling the 1980 gross value of agricultural and industrial output by the year 2000 has profound implications on the development of its mineral and energy sectors. Geologic and economic knowledge about the extent and availability of metal, non-metal, and energy resources in China must be the foundation of government-sponsored development policies. In addition, with the encouragement of international minerals co-operation in the forms of trade, investment, and joint-venture activities, China's mining industry leaders must remain aware of world market conditions. In China, the huge domestic demand for minerals and energy must be satisfied, suggesting expanded co-operation and development activities with the global minerals community. Such activities will be likely to require large capital expenditures for exploration, development, and the modernization of processing facilities.

SOCIALIST ECONOMIC GROWTH

The development of China's mineral and energy industries since 1949 has been largely affected/shaped by the nation's central planning process and chosen model of development. The future of these industries will be influenced by these factors as well, although the impacts of central planning may not be as strong if China's recent adoption and utilization of some market-oriented reforms continue uninterrupted.

In choosing a model of development in the early 1950s, the new government was concerned with advancing China's economy and administration and with initiating a series of progressive laws and regulations on trade, finance, and control of private enterprises. To meet the needs of a new nation striving for economic and political development, the government of China, in late 1952 and early 1953, sought an overall strategy or model for large-scale industrialization. At that time, two development models seemed to be relevant to Chinese development—the Soviet or Stalinist model, which existed under Stalin's rule between 1928 and 1953, and the Japanese model, which was in place from the Meiji Restoration until the Second World War (1868–1941).[3] Because of China's commitment to a centrally planned rather than a market economic system, Chinese leaders chose the Soviet model over the Japanese one (Cheng 1982: 258). Additionally, at that time, Chinese leaders expected significant Soviet economic aid, assistance, and support.

The Soviet model, when first implemented within the Soviet Union between 1928 and 1937, led to rapid industrialization, speedy growth, and marked transformation in the economy (Cheng 1982: 258).[4] The essential features of the model are: (1) a strong commitment to a high rate of economic growth;[5] (2) a very high rate of capital formation to sustain growth;[6] (3) an emphasis on the expansion of industry, particularly the capital goods industry; (4) concentration of investment in large plants with capital-intensive techniques; (5) a continuous funnelling of agricultural surplus into the industrial sector; and (6) a transformation in the agricultural sector to facilitate an intersector capital flow. Growth was largely of an extensive nature, based on huge increases in inputs, rather than of an intensive character based on higher efficiency with which new inputs were used. When the Soviet model was implemented within the Soviet Union, it entailed high costs, primarily because of low productivity growth resulting from irrational use of resources, rigidity of planning, weak workers' incentives, and other well-known defects of Soviet-type command planning. The huge input requirements of the industrial and defence sectors placed great pressure on the agricultural sector to support the industrial sector, which contributed to a long-term stagnation of agriculture.

Once the Soviet development model was chosen by the Chinese leaders, it was implemented in China in the form of a series of five-year plans (beginning in 1953), much like those in the former Soviet Union. However, despite the fact that a particular development model was chosen, China's economic history since 1953 is noted for its frequent fluctuations within the Soviet model framework. For nearly four decades no single economic strategy has lasted longer than six years. Oscillations in such strategies have resulted from and led to an interplay of many factors, including international relations, shifts between pragmatic and unreasonable ideological approaches, and changing economic conditions (Cheng 1982: 257).

Table 1.1 summarizes the evolution of China's development strategies between 1953 and 1995. The table describes briefly China's five-year plans implemented since 1953, and outlines the basic objectives of each plan. Simply stated, China's five-year plans provide an overall policy statement for a specific period indicating a particular direction the nation's economy and development should move towards. As the nation's five-year plans have been periodically interrupted by political changes and economic failures, column 2 in Table 1.1 lists the economic and political developments of these times.

Although Table 1.1 does not detail the content of China's development strategies, shifts in policies throughout the 1953 to 1995 period are apparent from the table. Heavy industry received a majority of investment in capital construction during each five-year plan period, and much of the emphasis throughout China's development evolution. The relative shares of investment to light industry and agriculture changed periodically, and by the Sixth Five-Year Plan period light industry was receiving a greater proportion of investment than agriculture. Agriculture received its largest relative share of investment during the Period of Readjustment (1963–5), as severe food shortages and general economic problems in the very early 1960s caused Chinese government officials to reconsider and adjust their development strategies.

The history of economic development in China from 1949 may be divided into four major periods (Hu, Zhang, and Liu 1984: 34–9):

TABLE 1.1. *Evolution of China's development strategies, 1953–1995*

Five-year plan	Major economic/ political developments	Basic objectives of five-year plans	Sectoral priorities[a]
First 1953–7	Accelerated pace of socialization of national economy and capital construction, 1955–6	Lay framework for socialist modernization	1. Heavy industry 2. Agriculture 3. Light industry
Second 1958–62	Great Leap Forward, 1958–60 People's Commune Movement, 1958–60	Accelerate economic development; realization of Second Five-Year Plan targets within three years; major decentralization of planning (1958–60)	1. Heavy industry 2. Agriculture 3. Light industry
	Recovery, 1961–2 Readjustment, 1963–5	Retrenchment and consolidation	1. Agriculture 2. Light industry 3. Heavy industry
Third 1966–70	Cultural Revolution, 1966–76	Rehabilitation, development of defence and rural small-scale industries[b]	1. Heavy industry 2. Agriculture 3. Light industry
Fourth 1971–5	Decentralization of planning, 1970; mini 'Great Leap Forward', 1971–2	Rehabilitation, development of defence and rural small-scale industries[b]	1. Heavy industry 2. Agriculture 3. Light industry
	Gang of Four, 1974–6		
Fifth 1976–80	Ten-Year Plan, 1976–85 Death of Mao Zedong, 1976 Fall of Gang of Four, 1976	Modernization of agriculture, industry, science and technology, and national defence, open-door framework; abandonment of Hua's overambitious ten-year plan	1. Heavy industry 2. Agriculture 3. Light industry
	Three-Year Adjustment,[c] 1979–81		
Sixth 1981–5	Accelerated economic reform and restructuring	Moderate growth targets, open-door policy, market reforms, reliance on foreign capital and technology	1. Heavy industry 2. Light industry 3. Agriculture
Seventh 1986–90	Three-Year Austerity Programme,[d] 1988–91	Moderate growth targets, open-door policy, and market reforms continue, reliance on technological exchanges and investment	1. Heavy industry 2. Light industry 3. Agriculture

TABLE 1.1. (*cont.*)

Five-year plan	Major economic/ political developments	Basic objectives of five-year plans	Sectoral priorities[a]
Eighth 1991–5[e]	Ten-Year Development Programme,[f] 1991–2000	Moderate, balanced, and sustained economic growth, sectoral decentralization and price reform, reliance on foreign assistance	1. Heavy industry[g] 2. Light industry 3. Agriculture

[a] Based on stated government policies and ambitions which changed through time with shifting development strategies and emphases. While agriculture was officially considered the starting-point of economic reconstruction during the Period of Recovery and Readjustment (1961–5), investment and pricing policies continued to favour growth of heavy industry. By 1991 China was seeking a more balanced growth of its national economy.

[b] Much of the contents of the Third, Fourth, and Fifth Five-Year Plans were not published until recently.

[c] In Dec. 1978 the Third Plenum of the 11th Chinese Communist Party Central Committee rejected Chairman Mao's general ideology and launched China into a new era by adopting 'open-door' economic reforms.

[d] An overheated economy, inflation, corruption, declining state control, and a growing trade deficit led to an adjustment or slowing down of economic reform in 1988 with the introduction of an austerity programme. During the year the Chinese government ordered most new industrial development projects halted. Student demonstrations in 1989 aggravated an already troubled economy.

[e] Incomplete.

[f] The Ten-Year Programme was adopted by the Fourth Session of the Seventh National People's Congress. It charts China's economic construction and development in the 1990s.

[g] The Eighth Five-Year Plan does seek to reverse the pattern of the 1980s when industry grew much more rapidly than other sectors, and substitute a more balanced pattern of growth.

Sources: *Beijing Review* (13–19 May 1991, 6–12 May 1991, and 15–21 Apr. 1991); Riskin (1987); Liu Suinian and Wu Qungan (1986); Su Kaiming (1986); Xinhua News Agency News Bulletin (1986); Foreign Language Press (1984); State Statistical Bureau (1960).

1. the basic completion of socialist economic transformation (1949–56)—China's socio-economic structure moved from one where an individual-oriented economy of peasants, handicrafts people, and other independent workers dominated the national economy to one where a public or state ownership by the whole people in the non-agricultural sectors and socialist collective ownership in agriculture dominated;
2. full-scale socialist construction (1956–66)—economic growth continued but on an erratic path after the Great Leap Forward (1958–60) and People's Commune Movement (1958) were instituted;
3. the Cultural Revolution (May 1966–October 1976)—marked as a period of economic chaos; and
4. the new 'historical' period since the end of the Cultural Revolution in October

1976—a restructuring of the economic managerial system in China and a readjustment of the national economy, including an opening to the Western world begun in late 1978.

The many socio-economic and political developments which occurred during the four periods of economic development in China brought about fundamental changes to the nation's economy. Economic growth was slowed considerably and even reversed in the early 1960s as a result of impractical and inefficient production processes encouraged by the Great Leap Forward and People's Commune Movement. Likewise, activities during the Cultural Revolution, many of which were unprecedented, resulted in periodic economic disorder. Post-Cultural Revolution policies, including the momentous open-door policy and market-oriented economic reforms adopted in late 1978 and implemented thereafter, strongly influenced the structure and development of China's economy in the 1980s. With the completion of the austerity measures of 1988–91, economic growth in 1992 began to accelerate once again.

Table 1.2 lists, in chronological order since 1953, the main indicators of China's economy—national income, gross value of agricultural and industrial output (GVAIO), population, labour force, total trade turnover, and the balance of financial revenues and expenditures. The National Income values reflect the sum of net output value of the five material-producing sectors of the Chinese economy—heavy and light industry; agriculture; construction; transport and communications; and commerce.[7] In contrast, GVAIO figures reflect the sum of gross values of all goods and services produced during a given period by the two largest sectors of the national economy—heavy and light industry and agriculture. Whereas National Income values account for input costs, there is no adjustment or netting out of input costs in the compilation of GVAIO figures. For this reason, GVAIO figures are often inflated, principally as a result of double-counting of intermediate inputs and the finished products. Intermediate inputs represent material goods needed to produce a finished product, such as pig iron in the production of steel.

Evident in Table 1.2 is the prominent structural change which occurred in China's economy since 1953. Agriculture dominated the economy in the early years, but industry soon moved ahead of agriculture as the largest contributor to GVAIO. As this was happening, the share of value that heavy industry gained relative to light industry was significant. This is partially a result of the emphasis on expansion of the heavy industry sector in six of eight five-year plans implemented in China, although, importantly, the size of the heavy industry sector may be overestimated because of high heavy industrial prices relative to other industrial goods prices. Total trade turnover in China since 1953 kept pace with economic expansion, as growth in exports earned the foreign exchange needed to purchase imported goods. Foreign trade has been considered a necessary component of economic growth, as the late Minister of Foreign Trade Yeh Chi-chung stated in a speech delivered at the Second Session of the First National People's Congress in

July 1955, 'Export is for the sake of import, which, in turn, is for the Socialist industrialization of our country.' Foreign trade and financial technology assistance remain a cornerstone of China's current economic development strategies.

Population in China increased an average of 1.8 per cent per year between 1953 and 1991. During this period China experienced a demographic transition from high to low birth and death rates, where the death rate fell rapidly in the 1950s to a low of nine deaths per thousand in the 1970s (Lavely 1989). This decline was interrupted by the death of an estimated 30 million Chinese during the three-year (1959–61) famine which followed the Great Leap Forward campaign. By 1991, the beginning of the Eighth Five-Year Plan period, China's population reached an estimated 1.158 billion persons.

INDUSTRIAL DEVELOPMENT IN CHINA

Since the founding of the People's Republic, China has developed an independent and extensive industrial base. Industry has always been deemed vital to China's economy and overall growth. New branches of industry have developed since 1949, including petrochemical and electronic industries, and many new technologies have been adopted and utilized. With the industries which produce minerals and energy already operating at very large scales, China's production of many commodities leads all other nations.

Chinese industry can be broadly divided into two sectors—heavy industry and light industry. The heavy industrial sector consists of the coal and coke-making, petroleum, power, metallurgy, machine-building, electronics, chemical, and building materials industries, whereas the light industrial sector comprises the pulp and paper-making, salt, sugar, cigarette, wine and liquor, bicycle, watch, sewing-machine, ceramics, tanning, textile, and leather and leather-making industries.

Before 1949, industries in China were unevenly developed and distributed, and light industry accounted for upwards of 70 per cent of the total value of industrial output. But by 1980 the distribution of Chinese industries was changed markedly, and light industry accounted for less than 50 per cent of China's gross industrial output value. The number of industrial enterprises with independent accounting systems reached 377,000 in 1980, and output of major industrial products was several times that of 1952. By 1992 there were nearly 420,000 industrial enterprises in China, the majority of which were collectively owned. Large and medium-sized industrial enterprises are all state-owned, and although they account for only a small proportion of the total number of industrial enterprises in the country, their share of total industrial output is over 50 per cent.

Table 1.3 presents average annual growth rates of industrial output value in China for each completed five-year planning period except the Seventh (1986–90).[8] In addition, rates for the 1953 to 1978 post-Liberation and 1979 to 1985 post-open-door eras are presented. The figures are disaggregated by major branch of industry, including the metallurgical, coal, and petroleum industries. These

TABLE 1.2. *Indicators of China's national economy, selected years,[a] 1953–1991*

Indicator	1953	1958	1963	1966	1971	1976	1981	1986	1991	Average annual growth rate	
										1953–91	1981–91
National Income (bn. 1980 yuan)	81.41	133.31	103.49	165.04	225.39	267.59	386.87	639.41	919.70	6.6	9.0
GVAIO (bn. 1980 yuan)	88.89	172.43	147.73	245.15	370.70	488.14	740.36	1,307.86	2,285.65	8.9	11.9
GVAO (bn. 1980 yuan)	86.60	107.33	93.75	125.5	143.95	164.02	204.61	294.75	372.74	3.9	6.2
GVIO (bn. 1980 yuan)	34.52	93.75	79.34	145.01	239.08	332.98	538.70	1,018.40	1,939.17	11.2	13.7
Population[b,c] (millions)	588.0	659.9	691.7	745.4	852.3	937.2	1,000.7	1,065.3	1,158.2	1.8	1.5
Labour-force[b] (millions)	213.6	266.0	266.4	298.1	356.2	388.3	432.8	512.8	583.6	2.7	3.0
General index of retail prices (1980 = 100)	78.6	82.6	97.6	91.1	88.8	90.0	102.4	125.6	209.4	—	—
Total trade turnover[c] (bn. current yuan)											
Exports	3.48	6.70	5.00	6.60	6.85	13.48	36.76	108.21	383.06	13.2	26.4
Imports	4.61	6.17	3.57	6.11	5.24	12.93	36.77	149.83	339.87	12.0	24.9

Balance of financial revenues and expenditures[f] (bn. current yuan)	0.28	−2.18	0.27	1.71	1.25	−2.96	−2.55	−7.05	−20.27	—	—

Notes: Figures are rounded. — Not applicable. GVAIO gross value of agricultural and industrial output GVAO gross value of agricultural output GVIO gross value of industrial output.

Since 1971, output values from rural industries have been grouped with industry instead of agriculture.

Real national income, GVAIO, GVAO, and GVIO figures were calculated by multiplying previous year's level by real growth rate.

[a] Beginning years of China's five-year or other major economic plans.

[b] Year-end figure.

[c] Figures refer to population of 30 provinces, autonomous regions, and municipalities on the mainland, including servicemen.

[d] Including list prices, negotiated prices, and market prices, similarly hereafter.

[e] Figures before 1980 are from Ministry of Foreign Trade, while figures since 1980 are from China's Customs Statistics.

[f] Domestic balance of financial revenues and expenditures. Financial revenues include the profits of state-owned enterprises and taxes received; financial expenditures include appropriations for capital construction and funds allocated to state-owned enterprises. Deficits generally are made up by increasing money supply.

Sources: State Statistical Bureau (1981–92), author's files.

TABLE 1.3. *Average annual increase rate of gross industrial output value by major branch of industry* (%)

Branch	Five-Year Plan							Post-Liberation (1953–78)	Post-Open Door (1979–85)
	First (1953–7)	Second (1958–62)	Adjustment (1963–5)	Third (1966–70)	Fourth (1971–5)	Fifth (1976–80)	Sixth (1981–5)		
GVIO	18.0	3.8	17.9	11.7	9.1	9.2	10.8	11.3	10.1
of which:									
Metallurgical industry	29.2	7.4	20.4	8.8	5.3	8.3	7.0	12.8	7.3
Power industry	20.4	20.7	12.8	11.6	10.8	8.6	7.6	14.7	7.7
Coal industry	17.1	11.8	0.1	8.8	5.4	4.8	5.6	9.2	3.6
Petroleum industry	32.7	22.2	27.4	18.5	14.6	7.0	5.1	20.9	4.8
Chemical industry	31.2	14.4	23.8	17.3	10.4	11.3	10.4	18.1	10.0
Machine-building industry	29.7	7.6	21.8	15.9	13.6	7.4	14.8	16.1	11.9
Building-materials industry	20.0	−4.5	30.0	9.4	11.5	12.7	12.4	11.8	11.3
Forest industry	13.7	−4.9	8.1	−1.9	7.5	7.2	4.8	4.4	5.1
Food industry	13.2	−1.7	11.4	2.4	8.4	8.0	9.2	6.3	9.4
Textile industry	8.6	−3.1	21.7	8.0	4.2	13.2	11.9	6.9	13.6
Paper-making industry	19.1	2.5	12.1	3.3	6.4	9.1	9.1	8.3	9.1

Notes: Figures in this table are at comparable prices which are obtained by multiplying the amount of products by their constant prices.

Beginning in 1985, the Chinese government amended its statistical reporting procedures to bring the nation's economic data into accordance with United Nations and World Bank reporting styles. Gross industrial output value was no longer disaggregated by industry as shown in the table, hence figures from the Seventh Five-Year Plan period are not available.

Source: State Statistical Bureau (1981–92 (1986, Beijing): 226).

three mining-related industries traditionally account for a relatively significant percentage of gross industrial output value. In 1988, for example, nearly 17 per cent of China's gross value of industrial output (GVIO) was attributed to activities of the metallurgical, coal, and petroleum industries.

China's GVIO grew at an average annual rate of 11.3 per cent between 1953 and 1978, a strong increase, particularly in the light of the economic turmoil created during the Great Leap Forward and Cultural Revolution movements. As shown in the table, however, despite the strong growth on average, there were periods of slow growth and even industrial decline, such as during the Second Five-Year Plan (1958–62). During this five-year lapse several light industries in China suffered losses in output, quality, and productivity, owing to the excessive emphasis on heavy industry (iron and steel) at that time, which claimed the major share of investment. It was not until the Period of Readjustment (1963–5) that the forest, food, textile, and paper-making industries rebounded.

The relatively high but erratic rate of growth of China's industry since 1953 was accomplished through increasing inputs of labour and capital, and characterized by declining prices in those branches of industry that produced heavy industrial goods—except for coal and timber—and by rising prices in light industry (Field 1986). Table 1.2 showed that China's total labour force increased between 1953 and 1991 at an average annual growth rate of 2.7 per cent. The amount of labour and capital invested in China's various industries through the years was largely a function of the economic objectives set forth by the five-year or annual economic plans and the political developments of the time.

During the First Five-Year Plan period, industrial growth was strong at an average annual rate of 18 per cent, reflecting a rebound from the Sino-Japanese war and subsequent civil war and the institution of systematic economic planning. Abundant Soviet capital and technical assistance facilitated industrial expansion during the First Five-Year Plan period. The overall rate of growth during the Second Five-Year Plan was strong at first, reflecting the Great Leap Forward initiatives, but nearly collapsed in later years owing to the exhaustion of material, financial and labour resources. Industrial growth rebounded once again during the Period of Readjustment (1963–5), but fell at the onset and during the middle of the subsequent Cultural Revolution period. During the post-1976 years, industry has been characterized by strong growth, averaging more than 10 per cent per year after the open-door reforms of 1978 were implemented.

PURPOSE AND SCOPE OF BOOK

Although today China's mining industry is very large by world standards, there are few comprehensive studies on the historical development of the industry, its overall impact on economic growth, and the procedure in which mineral and energy strategies are formulated and subsequenly implemented. There is indeed a general uncertainty about the entire industry. Until recently, there was difficulty in acquiring detailed and, at times, even general information about Chinese minerals and energy output, industrial organization, mining practices, ore grades, and long-term projections. After a virtual blackout of statistical data for nearly two decades, the Chinese government began releasing economic data, although primarily macroeconomic in scope, in the late 1970s. Minerals and energy data are becoming more readily available, though they vary, sometimes considerably, between Chinese sources. Such variation also commonly occurs between Chinese and Western sources of data.

This book is about the mineral and energy industries of China—how they have affected economic development since 1949, their present status, and the prospects and challenges ahead. It has been a dynamic industry since 1949, suffering large-scale losses in iron and steel output and quality with the Soviet withdrawal in 1960 and failures of the Great Leap Forward, rebounding with the discoveries of the Daqing and Shengli oil fields, suffering again with the onset of the Cultural Revolution, then advancing significantly in the late 1970s and early 1980s with recovery from the Cultural Revolution and open-door reform. The future of the industry is likely to be subject to dynamism, growth, decline, and turmoil. The dramatic ups and downs of the first four rounds of bidding for China's offshore oil are evidence of such dynamism.

The overall objective of this book is to define thoroughly the characteristics of the Chinese mining industry based on past and present performance. The many challenges that the industry will face in the years ahead will also be examined in the book. Other objectives include assessing the extent to which the structure and development of China's mineral and energy industries have been affected by a centrally planned economy, determining the present status and structure of the Chinese mining industry, examining the major problems currently plaguing the industry, and lastly, analysing the prospects for future growth of the industry. All minerals are included in the analysis, including fuel, metals, and non-metals.

The book consists of three parts. The first contains three chapters that present an examination of the mining development history of China and its relationship to the central planning process. Chapter 2 documents output levels of selected metals, non-metals, and mineral fuels in China for years after 1952, to gain an understanding about the development of the Chinese mining industry. China's GVIO since 1952 is also analysed and compared with changing levels of mining output. In this chapter the evolution of China's mining industry is disaggregated into four major periods, where each period is noted for either stability or instability of output. Because China's mining industry did not develop in isolation from other sectors, its interrelationships with such sectors as agriculture are briefly reviewed in Chapter 2. Today, an important linkage exists between agriculture and mining in China, namely the peasant farmer, who is increasingly entering a more profitable profession—mining.

The importance of central planning to minerals and energy development in China is the focus of Chapter 3. Every five-year and other major political plan is analysed in the chapter in terms of policies which affected heavy industry and, in turn, China's mineral and energy industries. The impacts of those policies on production of metals, non-metals, and mineral fuels is thoroughly documented. The following specific periods of China's history are evaluated—the Period of Rehabilitation (1949–52), the First Five-Year Plan (1953–7), the Second Five-Year Plan (1958–62), the Great Leap Forward (1958–60), the People's Commune Movement (1958–60), the Period of Recovery and Readjustment (1961–5), the Third Five-Year Plan (1966–70), the Cultural Revolution (1966–76), the Fourth Five-Year Plan (1971–5), the Fifth Five-Year Plan (1976–80), the Ten-Year Plan (1976–85), the Sixth Five-Year Plan (1981–5), the Seventh Five-Year Plan (1986–90), the Eighth Five-Year Plan (1991–5), and the Ten-Year Development Programme (1991–2000). Because of the devastating economic impacts of the Great Leap Forward, the People's Commune Movement, and the Cultural Revolution, these three political movements are examined in detail in terms of their adverse impacts on overall mining development in China.

The second part of the book, comprising Chapters 4, 5, and 6, describes the economic characteristics of the Chinese mining industry, including its interrelationships with other sectors of the national economy. Chapters 4 and 5 evaluate the present status of China's mining industry in terms of geologic conditions, minerals endowment, production, consumption, and trade. Chapter 4 presents the current organizational structure of the entire Chinese mining industry, which was thoroughly revised in 1988 in an attempt to streamline activities, initiate more efficient operations, provide more incentives, and attract foreign investor interest. Chapter 5's analysis of China's minerals and energy trade focuses on recent activities, but export and import figures are presented covering selected years from 1970. Other forms of international minerals co-operation, including direct foreign investment and joint-venture activities, are also described in Chapter 5.

To comprehend the economic impact of China's metallurgical, coal, and petroleum industries on other producing and consuming sectors in the nation's economy, an input–output analysis of the Chinese economy is presented in Chapter 6. An important question about the three mineral-related industries in China is how they directly and indirectly influence the output, income, and employment of other sectors in China's economy, especially those sectors which supply inputs directly to the mineral-related industries. The input–output technique quantitatively and systematically describes China's economy as a system of interdependent sectors, and demonstrates how an initial economic stimulus or change in China's minerals sector can produce widespread secondary and tertiary effects which are felt throughout the nation's economy. This phase of the analysis focuses on the economic impact of China's mineral and energy industries on national output and income. The input–output modelling is intended to provide a snapshot view of the interrelationships between mining and the Chinese economy.

Examination of the historical development of the Chinese mining industry reveals patterns which shed light on the possible future course of the industry. Part III of this book examines the long-term outlook of mining development in China. Chapter 7 evaluates the various options open to the Chinese government to stimulate future growth of mining, along with major government policies for long-term exploration and development activities. Several problems which may hinder future minerals and energy development in China are outlined, including limited foreign investment, the local–state dichotomy of interests, an irrational pricing system, transportation bottlenecks, environmental degradation, and inefficient mining by peasants (farmers). Unless some of these problems are addressed in earnest by the Chinese leadership, a conflict may arise between the nation's ambitious long-term economic targets and resource availability. International perceptions and misconceptions about China's mining industry are also described in Chapter 7, in an attempt to understand changing levels of foreign investor interest in the nation's mining industry.

The summary and conclusions of the book are presented in Chapter 8. In it, the principal qualities which characterize the Chinese mining industry are again restated, and some of the patterns of development of the industry are reviewed. China's mineral and energy industries are critical to the overall economy of the nation, as evidenced in most of China's plans, where their growth was considered a priority. With the current modernization programme under way, China's mining industry is continuing to receive strong support and financial backing, but problems such as a general lack of hard currency and rising imports are afflicting the industry. The future outlook of the industry is promising, though a degree of uncertainty looms over long-term forecasts.

NOTES

1. For purposes of this book, the mining industry of China is defined as including the mineral and energy industries, where 'mineral' refers to metals and non-metals and 'energy' denotes petroleum, coal, natural gas, and their products. The mining industry comprises such activities as exploration, development, and mining of metallic and non-metallic minerals, petroleum, natural gas, and coal as well as mining-specific infrastructure. Also included are the smelting, refining, primary rolling, and fabricating of non-ferrous basic metals, and downstream petroleum refining activities.
2. With the 1992 production level, China ranks first in world production of hard coal and sixth in petroleum output: British Petroleum Company 1984–93 (June 1993). Note that the coal production figures refer to raw coal output; in terms of clean or washed coal

production, China ranks second worldwide behind the United States. In terms of energy content, China produced 543.7m. tonnes oil equivalent of coal in 1992, compared to 538.3m. tonnes oil equivalent in the United States.

3. These two models seemed to be relevant in terms of promoting large-scale industrialization, which was desired by the Chinese leaders at that time.
4. For a detailed discussion of the Stalin development model, see Yeh (1967) and Cohn (1970), ch. 4.
5. Between 1928 and 1937, the annual growth rate of Soviet GNP (in constant prices) was greater than 8%.
6. During the early years of Soviet industrialization (1928–37), the rate of capital formation was as high as 20–25%.
7. Unlike Western GNP figures, there is no accounting of the services sector and depreciation of assets in China's National Income values. It has been estimated by East–West Center officials that if the services sector and depreciation were accounted for, China's National Income figures of recent years would have been approximately 13% higher.
8. Measured using the factory method, which is a means of computing GVIO in China based upon a unit of industrial enterprise—the basic unit of the nation's economic system. According to the factory method, GVIO is based on the values of finished products of the production processes of all industrial enterprises and does not include materials consumed within these processes. That is, GVIO does not represent a sum of values of products of each workshop within each factory (intermediate goods), but rather the sum of the values of final products.

2
A Historical Perspective on Mining and Economic Change

GEOLOGICAL EXPLORATION IN CHINA

Historical records suggest that the gathering of geologic information in China began more than 2,000 years ago. Modern-day geological work in the nation started in the early part of this century, but it progressed rather slowly. Before the founding of the People's Republic, there were only 800 people engaged in geological work, and only 200 trained geologists among them (Wang Xionglin 1988).

Table 2.1 lists the number of personnel engaged in geological exploration in China between 1950 and 1991. These figures include personnel from the Ministries of Geology and Mineral Resources and Metallurgical Industry, and the former Ministries of Coal Industry, Petroleum Industry, and Construction Materials. Approximately a million persons were engaged in geological and minerals exploration for several Chinese ministries in 1991. This figure includes all employees—technical, non-technical, and assistants. According to employment figures in Table 2.1 the average rate of growth per annum of geological personnel between 1950 and 1991 was a remarkable 18.3 per cent, representing an apparent strong commitment by the Chinese government to geological exploration, particularly in the early 1950s, and the 1970s on. Total employment declined or remained steady during the 1960s, as a direct result of the Cultural Revolution.

Technical staff engaged in geological and minerals exploration includes geoscientists and mining personnel—these are the types of employees normally counted as workers in geological surveys and other similar organizations in Western nations. As a result, it may be more appropriate to discuss technical staff figures of China, particularly when comparing and contrasting with employment figures of Western nations. At present, an estimated 159,000 technicians are involved in geological exploration activities in China. This compares to approximately 9,500 full-time employees at the US Geological Survey, the largest government-funded geological research institution within the United States.

The figures in Table 2.1 can be disaggregated by planning period to reveal historical changes in the average annual growth rates of geological personnel in China. They are: (1) 223.6 per cent—Period of Rehabilitation (1950–2); (2) 45.2 per cent—First Five-Year Plan (1953–7); (3) −10.5 per cent—Second Five-Year Plan (1958–62); (4) −4.2 per cent—Period of Recovery and Readjustment (1961–5); (5) 7.3 per cent—Third Five-Year Plan (1966–70); (6) 4.3 per cent—

Fourth Five-Year Plan (1971–5); (7) 5.1 per cent—Fifth Five-Year Plan (1976–80); (8) 1.2 per cent—Sixth Five-Year Plan (1981–5); and (9) −1.2 per cent—Seventh Five-Year Plan (1986–90). As already mentioned, rapid growth of personnel occurred in the early 1950s, while the growth rate slowed considerably during the Cultural Revolution. Because of the increasing demands for mineral and energy resources and products in the 1980s, the Chinese government gradually increased the technical staff of geoscientists, engineers, chemists, and machinery operators. Renewed emphasis on geological exploration activities in the 1990s may lead to an increase in the number of exploration personnel during the decade.

During the period between 1949 and 1991, many goals were achieved within China's mining industry in respect to geological exploration and development.

TABLE 2.1. *Growth of personnel*[a] *engaged in geological exploration*[b] *in China* ('000)

	Total (end of year)	Including	
		Technical staff[c,d]	Workers and apprentices[d]
1950	1.7	0.3	1.2
1951	4.1	0.5	2.6
1952	17.8	2.4	10.4
1953	64.3	7.7	37.1
1954	97.8	13.2	49.7
1955	129.4	18.9	69.4
1956	271.6	35.3	150.1
1957	286.2	41.4	158.8
1958	455.9	47.9	273.3
1959	523.7	67.7	342.1
1960	618.8	74.2	389.0
1961	458.8	76.0	249.5
1962	292.6	65.5	148.2
1963	310.8	70.8	161.1
1964	356.0	73.6	183.2
1965	386.9	80.1	202.9
1966	420.1	80.2	234.7
1967	425.2	81.3	240.9
1968	429.6	80.7	249.2
1969	482.7	78.8	310.1
1970	556.7	81.6	364.2
1971	637.8	87.4	418.4
1972	643.1	90.0	393.6
1973	655.1	91.5	403.1
1974	683.6	90.7	413.3
1975	754.3	95.0	462.2

TABLE 2.1. (*cont.*)

	Total (end of year)	Including	
		Technical staff[c,d]	Workers and apprentices[d]
1976	825.5	103.0	510.0
1977	869.4	112.9	535.3
1978	939.4	123.0	583.9
1979	984.4	124.1	599.8
1980	1,006.3	130.1	599.5
1981	1,012.3	133.6	583.6
1982	1,012.3	141.0	566.0
1983	1,030.8	146.0	571.0
1984	1,052.3	151.1	579.2
1985	1,060.0	154.0	578.0
1986	1,050.0	154.0	567.0
1987	1,070.0	159.0	573.0
1988	1,070.0	162.0	567.0
1989	1,040.0	159.0	546.0
1990	1,000.0	156.0	520.0
1991	1,000.0	159.0	533.0

[a] Includes personnel from the Ministries of Geology and Mineral Resources, Metallurgical Industry, Coal Industry, Petroleum Industry, Chemical Industry, and Construction Materials Industry. In 1988, during a government reorganization in China, the coal, petroleum, and nuclear industries, as well as part of the Ministry of Water Resources and Electric Power, merged to become the Ministry of Energy.

[b] Geological exploration encompasses all phases of exploration including prospecting, minerals exploration, and estimation.

[c] Technical staff includes such personnel as professional geoscientists, engineers, chemists, and machinery operators.

[d] 1985–91 technical staff and workers and apprentices figures based on author's estimates.

Sources: Dizhi Chubanshe (1985); and State Statistical Bureau (1981–92 (1992, Beijing): 104).

Some of these achievements are as follows (*Beijing Review*, 4–10 Mar. 1991; Ministry of Geology and Mineral Resources 1991*a*, *b*; *China Daily*, 27 Nov. 1989; Xinhua News Service, 30 Aug. 1989; Wang Xionglin 1988; Dizhi Chubanshe 1985; Zhou Guojun 1985):

- The number of types of minerals with verified reserves increased from 18 (pre-Liberation) to 162 (1989).
- Mineral deposit localities grew from 5,123 (pre-Liberation) to nearly 200,000 (1990). Of these 200,000 localities, more than three-quarters have verified mineral reserves.

- The output of the ten most commonly used non-ferrous metals in China reached 2.52m. tonnes in 1991, a 33.9-fold increase over 1952, the year of an overall economic rehabilitation.
- Total exploration drilling reached 25m. metres in 1988, compared to 170,000 metres in 1949. Regional geological maps at a scale of 1:1,000,000 covering the entire country have been completed, while regional geological and hydrogeological maps at a scale of 1:200,000 covering two-thirds of the country have also been finished.
- Prior to 1949, China was regarded as an oil-poor country. Since the 1950s, China has developed more than 18 onshore oil and gas fields and seven offshore oil fields, yielding an annual oil output of nearly 140m. tonnes. Daqing, Shengli, and Liaohe oil fields each produce more than 13.5m. tonnes annually.
- Employment in the minerals industry engaged in geological exploration increased from 800 (pre-Liberation) to 1 million (1991). The technical staff in 1990 represented about 16 per cent of total personnel.

A major accomplishment of the geological community of China since 1949 was the noticeable addition to verified mineral and energy reserves in the nation. Figure 2.1 graphically illustrates the percentage increase in discovered reserves of major resources in China, as reported in Chinese documents. More recent reserve figures are not available. The minerals depicted include troilite (sulphur ore), petroleum, gold, bauxite, phosphate rock, iron ore, coal, and copper. Reserves in China represent that portion of the resource base that is technologically and economically recoverable given today's market conditions. Because of the dynamic nature of reserves, estimates of resource endowment can change given additional methods of discovery, evaluation and processing, and changing prices. As a consequence, long-term changes in reserve quantities provide less than conclusive estimates of total resource endowment, but an accurate indication of the pace of reserve discoveries through time.

As illustrated in Figure 2.1, reserves of petroleum in China increased nearly 250 per cent between the First Five-Year Plan period and 1983. Today, China's proved crude oil reserves amount to nearly 24bn. barrels, while petroleum reserve amounts in 1949 were negligible (British Petroleum Company 1984–92 (June 1992)). Nearly half of all known Chinese oil originally in place was discovered in one of three years—1958, 1959, and 1962. Additions to reserves recorded in recent years have, however, been substantial.

With the continuing discoveries and verification of mineral and energy reserves in China since 1949, increased development of these commodities was possible. However, for some metals in particular, such as iron ore and copper, the newly found reserves could not keep pace with increased demand in China, necessitating imports. For others, like gold, domestic production continued to grow rapidly in China while reserves were being added.

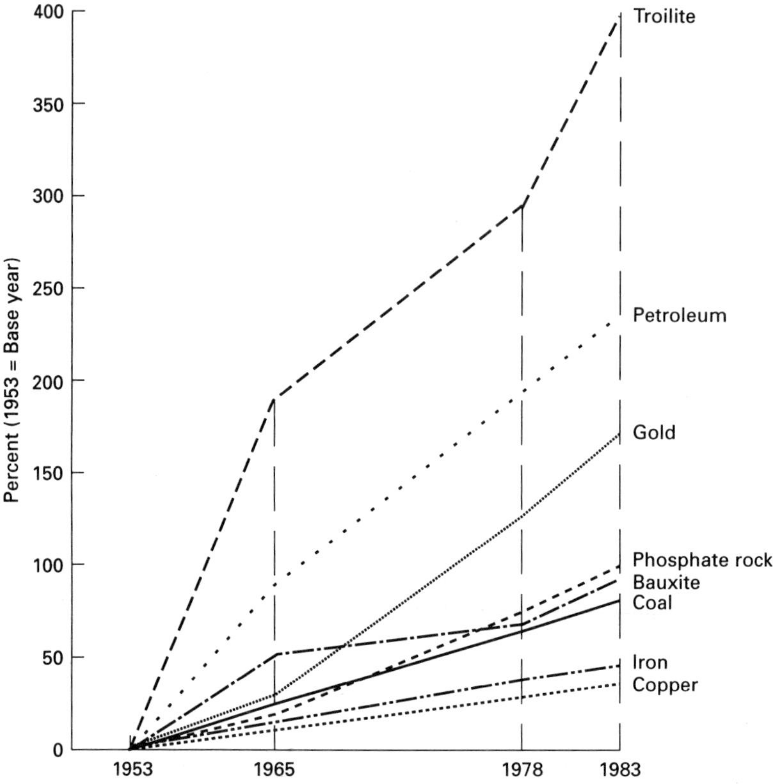

FIG. 2.1. Increases in discovered mineral and energy reserves of China

INVESTMENT IN THE MINING INDUSTRY

One way of characterizing a government's commitment to the development of its national mining industry is by assessing the growth (or decline) of capital investment levels over time. During the 39-year period between 1953 and 1991, approximately 33.5 per cent of China's state investment in the capital construction of heavy and light industries went to the metallurgical, coal, and petroleum industries. This represents a strong commitment on the part of the Chinese government to develop an extensive mining base. As highlighted in Chapter 1, the metallurgical, coal, and petroleum industries have traditionally represented a major contribution in value to industry.

Table 2.2 lists historical values of investment in capital construction by the Chinese government into the mineral-related branches of industry. Capital construction refers to construction, expansion, transformation, and restoration projects of all sectors of the national economy, including the construction of mines and factories. The investment in capital construction shows the amount of work done in capital

TABLE 2.2. *Investment in capital construction by branch of industry*[a] (bn. yuan/%)

Years (Period)	Branches			
	Total state industrial investment[b,c]	Metallurgical industry	Coal industry	Petroleum industry
1953–7 (First)	25.3 (100)	4.7 (18.6)	3.0 (11.7)	1.2 (4.7)
1958–62 (Second)	72.8 (100)	16.9 (23.2)	8.7 (11.9)	2.5 (3.4)
1963–5 (Adjustment)	21.0 (100)	3.4 (16.2)	2.5 (12.0)	1.6 (7.8)
1966–70 (Third)	54.2 (100)	9.9 (18.2)	4.7 (8.6)	3.9 (7.2)
1971–5 (Fourth)	97.8 (100)	17.3 (17.7)	9.1 (9.3)	8.9 (9.1)
1976–80 (Fifth)	123.2 (100)	19.0 (15.4)	13.6 (11.1)	13.1 (10.7)
1981–5 (Sixth)	154.7 (100)	21.2 (13.7)	20.3 (13.1)	14.6 (9.4)
1986–90 (Seventh)	378.6 (100)	40.4 (10.7)	35.2 (9.3)	37.9 (10.0)
of which:				
1986	53.2 (100)	5.4 (10.2)	5.8 (0.9)	3.9 (7.3)
1987	68.3 (100)	7.9 (11.6)	6.0 (8.8)	5.9 (8.6)
1988	79.6 (100)	9.5 (11.9)	6.4 (8.0)	8.6 (10.8)
1989	82.2 (100)	9.0 (10.9)	7.1 (8.6)	9.4 (11.4)
1990	95.3 (100)	8.6 (9.0)	9.9 (10.4)	10.1 (10.6)
1991 (Eighth)	114.7 (100)	12.5 (10.9)	11.7 (10.2)	12.6 (11.0)
TOTAL (1953–91)	1,042.3 (100)	145.3 (13.9)	108.7 (10.4)	96.3 (9.2)

Note: Figures are rounded.

[a] Refers to state-owned industrial enterprises.

[b] Includes the following heavy and light industries—metallurgical, power, coal, petroleum, chemical, machine-building, forest, building materials, textile, food, and paper-making.

[c] Beginning in 1985, the Chinese government amended its statistical reporting procedures to bring the nation's economic data into accordance with United Nations and World Bank reporting styles. Investment in capital construction figures for the metallurgical industry are now disaggregated as ferrous metals mining and dressing, non-ferrous metals mining and dressing, smelting and processing of ferrous metals, and smelting and processing of non-ferrous metals; coal industry as coal-mining and dressing; and petroleum industry as petroleum and natural gas extraction and petroleum processing.

Source: State Statistical Bureau (1981–92 (1986–92, Beijing)).

construction in money terms, or the scale and progress of capital construction during a certain period of time. Investment in capital construction is one component of total investment in fixed assets, with the other component being investment in technical renovation and transformation. Cumulative government funds invested for each five-year planning period are shown in Table 2.2, along with year investment values for 1986 through 1990, as well as 1991. To provide for rapid comparison between the levels of investment into each of the three mineral-related industries, percentages of total investment are also provided in Table 2.2.

Throughout the past four decades, the relative levels of investment devoted to the metallurgical, coal, and petroleum industries varied considerably, as shown in the table. Because of the early emphasis on industrialization and the building of capital-intensive industries, capital investment in the metallurgical industry was significantly larger than in the coal and petroleum industries during the first two five-year plans in China. Nearly 19 per cent (4.7bn. yuan) of China's investment in capital construction was geared for the metallurgical industry during the First Five-Year Plan, as compared to 11.7 per cent (3.0bn. yuan) to the coal industry and 4.7 per cent (1.2bn. yuan) to the petroleum industry. Until 1983 the metallurgical industry was made up of both steel and non-ferrous segments, and much of the investment into the industry traditionally went to the steel component.

With the Great Leap Forward begun in 1958, Chairman Mao's call for accelerated industrialization implied rapid growth of China's iron and steel industry. To accomplish this, investment into the metallurgical industry was increased rapidly during the Second Five-Year Plan period, as 23.2 per cent of China's total investment in construction of industry was designated for metallurgy development. However, with the Soviet withdrawal of technical assistance in 1960 and more importantly the huge losses incurred during the Great Leap Forward, investment in general and in the iron and steel industry in particular declined considerably.

As China's development priorities shifted through the years, capital investment levels by industry changed accordingly. By the mid-1980s, for example, the relative share of investment into China's coal industry was greater than that of both the metallurgical and the petroleum industries. This reflected efforts at that time by the Chinese government to step up construction of large- and medium-sized coal-mines to improve the geographical distribution of China's coal industry and to alleviate the shortage of coal in east, north-east, and south-central China. In 1988, the share of investment into the petroleum industry of China rebounded to become larger than that into the nation's coal industry, in an attempt to stem the decline in output at virtually all oil fields in China. Government policies which influenced the amount of investment geared for China's various mineral-related industries will be discussed further in Chapter 3, which will focus on the impacts of China's five-year plans on minerals and energy output.

Increased investment into a particular industry in China has generally been done to boost the size, output, and efficiency of the industry. It is difficult to assess the return on capital investment or, more specifically, the efficiency of capital invested in China's mining industry because of the country's highly distorted price system and ambiguities in Chinese accounting categories and conventions. As noted above, levels of capital invested in construction of China's coal industry increased significantly relative to other industries during the early 1980s. The extent to which this boost in investment contributed to output growth of coal and improved efficiency in the industry is hard to quantify accurately. Historical output trends of coal and other commodities is the subject of the next section.

HISTORICAL PRODUCTION TRENDS

Historically minerals and energy production in China has been subject to erratic behaviour, though generally increasing in volume since 1949. During the past four decades, minerals and energy output has grown from a relatively small base to very significant quantities. China's total production in 1991 was estimated at more than 40 times that in the early 1950s. Output continues to rise today. Before the founding of the People's Republic, there was little opportunity to develop the country's minerals base systematically. Wars and foreign control over parts of China severely plagued industrial development.

Beginning in 1952, mining output in China fluctuated with changing political and economic conditions, different strategies of capital investment, and new discoveries of mineral or energy deposits. As a means of depicting relative output change over time, Figure 2.2 illustrates varying index levels (1952 = 100) of coal, cement, steel (rolled), and salt production in China between 1952 and 1991. Output levels were indexed to provide a simple means of comparison of production of the four commodities. Index numbers of China's GVIO, or overall industrial production value, are also shown in the figure.

Over a period of time, the indices of output of coal, cement, steel (rolled), and salt generally reflect trends in production of other mineral and energy commodities, although the rate of growth of crude oil output was extremely rapid once production at Daqing began in earnest. Indeed, by 1963, the exploitation of Daqing helped China achieve self-sufficiency in most petroleum products, although the extent of the self-reliance at this time has been disputed. Unlike many other commodities, production of oil in China was not severely hit by detrimental political developments owing to its importance to the nation's overall economy and self-sufficiency motives.

Considering individual commodities of China's mining industry represented in Figure 2.2, the rolled steel index showed the most substantial gain between 1952 and 1960—a 948 per cent increase. Coal output during the period, although rising at a slower rate, grew more than 501 per cent. The cement and salt indices rose at slower rates between 1952 and 1960, with the index for salt having shown a 160 per cent increase.

As indicated in Figure 2.2, 1960 marks the beginning of a decline in output of several mineral and energy commodities in China, despite the ambitious Great Leap Forward programme. In terms of China's mining output, the years immediately following 1960 were ones of extreme volatility and general lack of direction. Although progress was occurring in the nation's drive to develop a petroleum industry, mining and industrial activities were slowed during much of the 1960s and early 1970s. The discovery of the Shengli oil field in 1962 (China's second largest producer), initiation of large-scale production at Daqing in 1963, and opening up of the Dagang oil and gas field in 1967 were all promising developments in China's oil industry, although it too was adversely affected by the Great

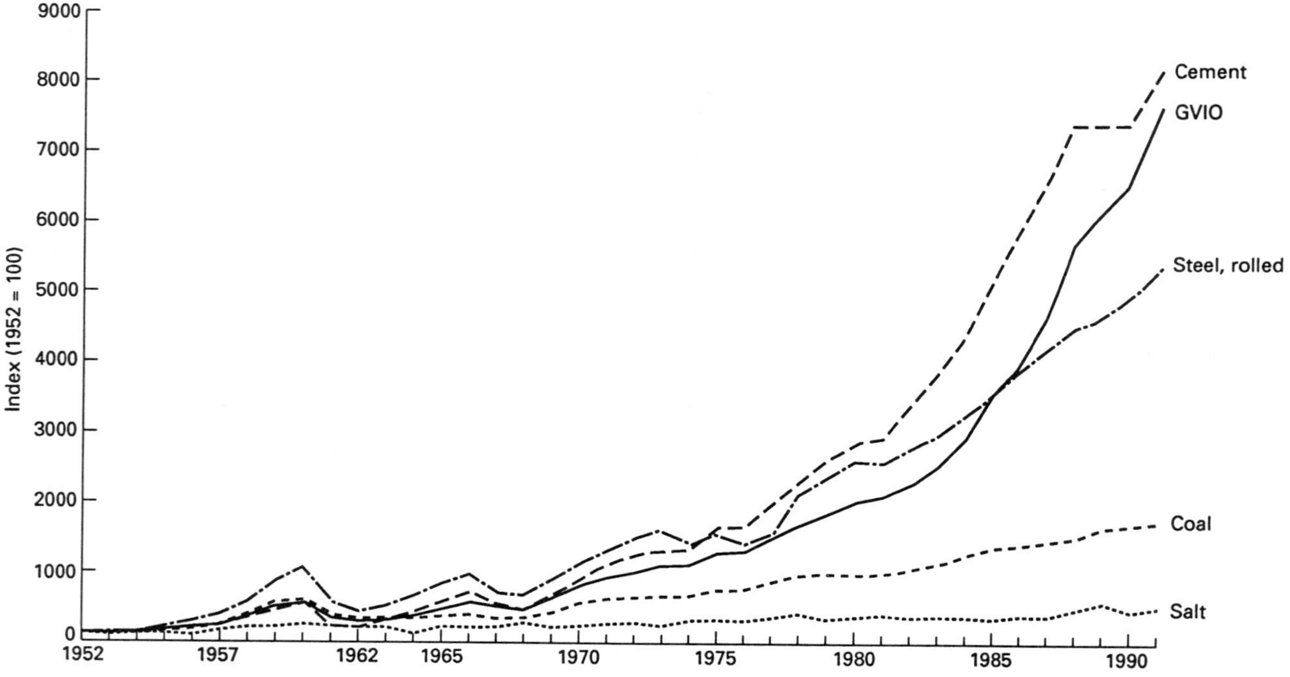

FIG. 2.2. Indices of physical volume of minerals and energy output in China, 1952–1991

Leap Forward and, in part, by the withdrawal of Soviet financial and technical assistance.

Altogether, there have been four major periods in the evolution of China's mining industry beginning in 1952, where a period is designated as such if there was a distinct and continuous long-term trend in output of major mineral and energy commodities. Defining such periods for an entire mining industry is admittedly difficult owing to the heterogeneous nature of the metals, non-metals, and mineral fuel industries. None the less, based on China's published production amounts for major mineral and energy commodities between 1952 and the present, four broad periods are defined in Table 2.3.

As already indicated, growth in output of metals (particularly tungsten and rolled steel), non-metals (salt), and coal in China was positive and rapid after rehabilitation and reconstruction of the economy was mostly complete in 1952. This trend lasted until about 1960, when industrial and mining output began to drop sharply. Period II, which lasted from 1960 to about 1971, was noted for fluctuating output levels throughout China's mining industry. Although by the mid-1960s minerals and energy output had begun to rebound from their recent low levels, the advent of the Cultural Revolution again slowed down industrial activities. Even crude oil output, which had steadily increased from 1960, declined for the first time between 1966 and 1967, from 14.6m. to 13.9m. tonnes. Armed clashes between industrial workers and Red Guard students caused major disruptions in China's industrial and mining activities. The coal and metallurgical industries of China were particularly disturbed by these activities.

By the late 1960s and early 1970s China's mining industry had recovered from the Cultural Revolution and long-lasting effects of the Great Leap Forward. In 1971 output of coal, rolled steel, and salt had reached or exceeded 1960 levels. Mining and industrial activities were essentially normalized at the start of the Fourth Five-Year Plan (1971) and government policies stressed the expansion of the nation's heavy industries, including iron and steel.

The remaining two periods in the evolution of China's mining industry occurred between 1971 and 1980, and from 1980 to the present. The latter period has been distinguished by open-door reform and the adoption of market-oriented measures. Decentralization has reduced central government control over mining output, distribution, and prices, although Beijing officials still largely supervise capital construction and investment activities. With the Cultural Revolution behind, high priority was given to expansion and development of China's energy industries, and renovation of the country's ageing iron and steel facilities.

In looking back at the overall evolution of the Chinese mining industry since 1952, it is apparent that three commodities drove the industry during much of the nation's development—coal, steel (rolled and other products), and petroleum (and petroleum products). Two non-metals, namely salt and cement, have also contributed greatly to China's economic and industrial development. China has

TABLE 2.3. *Major periods[a] in the evolution of China's mining industry*

Period[a]	Years	Major commodities produced[b]	Principal characteristics
Period I	1952–60	Tungsten, rolled steel, salt, and coal	Mining output increased several-fold during the period with large-scale construction of new mine plants. Strategies of accelerated industrialization led to dramatic rise in steel output.
Period II	1960–71	Rolled steel, salt, and coal	Economic chaos created by Great Leap Forward, Soviet withdrawal, and Cultural Revolution caused significant downturn in the nation's mining industry. Between political events industry rebounded. Crude oil output rose 7.6 times during the period.
Period III	1971–80	Rolled steel, salt, cement, and refined petroleum products industries	Chinese mining industry steadily progressed with advent of Fourth Five-Year Plan and simultaneous expansion of large and small industries. China achieved the status of a major world producer of crude oil by middle of period.
Period IV	1980–present[c]	Rolled steel, cement, and refined petroleum products	Nationwide economic reforms altering the character of the Chinese mining industry. Foreign investment, technology, and equipment being used to intensify exploration and development activities. Government restructuring of mining industry in 1988 changes the duties and responsibilities of the major players. Rate of growth of industry begins to slow towards end of period.

[a] Periods characterized by continued stability or instability of output.
[b] In terms of output value, linkage to overall economy, and revenues generated through taxes or foreign sales.
[c] Refers to 1980 to 1992.

consistently been one of the world's leading producers of salt since the 1950s. Salt production, which takes place in the majority of the Chinese provinces, has historically yielded huge revenues to China's government in the form of taxes and export earnings. Salt continued to be China's most valuable domestically produced non-metal until the mid-1970s, when the Chinese cement industry had finally reached world-class size. Accelerated rural and urban construction in the 1980s has contributed to China becoming the leading global producer of cement today.

Including years prior to 1952, coal has represented the greatest portion of China's minerals and energy wealth and the backbone of the nation's energy economy.[1] Steel and steel products became vitally important to China's industrial development when the drive towards industrialization was begun in the 1950s. Between 1952 and 1957 rolled steel output increased fourfold. Other metals of substantial economic importance to China have included tungsten, tin, and gold. Today's most important commodity produced in China in terms of output value and foreign sales is petroleum (with petroleum products). China became a major world producer of petroleum in the mid-1970s, and by the mid-1980s a major exporter of the energy commodity to world markets. The nation's refining abilities expanded greatly in the 1960s, and at an even faster pace during the 1970s. Petroleum exports have helped China accumulate the hard currency needed to modernize its petroleum facilities and intensify exploration and development.

As China's mining industry evolved through four major periods, many characteristics of the industry changed accordingly. Table 2.4 documents changes of four such characteristics, namely, output value, contribution of mining to GVIO, overall labour productivity, and tonnage of mining equipment used.

It is difficult to determine mining output values in China accurately as the nation boasts an independent and complicated pricing system far removed from international market forces. In addition, prices of most mineral and energy commodities in the country have never been published or revealed, especially during the 1960s when a virtual blackout on the release of all economic and statistical data existed. Prices in China do not reflect scarcity of resource nor supply–demand interactions as they do in most market-based economies. Consequently, the output values listed in Table 2.4 should be evaluated with caution, and readers should consider the magnitude of the changes in value, rather than the absolute values themselves.

As seen, by 1984 the estimated value of minerals and energy output in China had reached more than $50bn.,[2] making it the fourth most valuable mining industry in the world behind the former Soviet Union, the United States, and Saudi Arabia (Callot 1985). This represents remarkable growth of value since 1950, by an estimated 45 times. China's mining industry in 1983 accounted for about 6 per cent of global output value, whereas it represented just more than 1 per cent in 1950. During that year China's output value ranked sixteenth worldwide,

TABLE 2.4. *Historical mining and the Chinese economy, 1952–1984*

Indicator	1952	1957	1965	1978	1984
Minerals and energy output value ($USm. of 1983)	1,160[a]	NA	NA	46,200	53,300[b]
Total energy production[c] (m. tonnes of standard fuel)	48.7	98.6	188.2	627.7	778.5
Contribution of mining-related activities to GVIO (%)[d]					
All branches	NA	12.7	16.2	16.7	15.6[e]
Metallurgical industry	NA	8.8	10.2	8.7	8.2
Coal industry	NA	3.1	2.4	2.4	2.6
Petroleum industry	NA	0.8	3.6	5.6	4.8
Labour productivity in value terms (yuan/person/year)[e]					
All branches	4,184	6,362	8,979	11,130	14,070
Metallurgical industry	5,581	11,620	16,917	13,038	16,048
Coal industry	3,183	4,799	3,149	3,526	3,534
Petroleum industry	9,766	17,084	31,028	60,965	51,380
Mining equipment (tonnes)	1,800	52,900	40,000	242,900	258,100

Notes: NA not available. Figures are rounded.

Mineral and energy output values are crude approximations owing to an independent pricing system in China. For each mineral the value of the form in which it is normally marketed has been used, which in many cases will be its value after beneficiation. Tonnages used to calculate the values were taken from Chinese sources wherever available or from publications of the World Bureau of Metal Statistics, US Department of Energy, or US Bureau of Mines.

[a] 1950 est. value.

[b] 1983 est. value.

[c] Excludes bio-energy, solar, geothermal, and nuclear energy.

[d] Mining-related activities refer to extractive industries of metallurgy, coal, and petroleum.

[e] At 1980 constant prices.

Sources: Output values—Callot (1985); other indicators—State Statistical Bureau (1981–92 (1985–8)).

indicating that its mining industry expanded at a faster pace relative to many others during the past four decades.

As China's mining industry evolved through the years, varying growth rates of production of individual commodities meant that their respective output values changed accordingly. Table 2.3 listed the principal commodities produced in each of the four historical periods. As a group, the mineral fuels represented nearly half of China's mining output value by the early 1960s, and nearly two-thirds of the value a decade later (Dorian 1987). In 1974, China's Xinhua News Agency revealed just how greatly the country's petroleum industry had expanded by that year when it announced that Daqing had increased its production capacity fivefold since 1966, and output from Shengli, Dagang, and Yumen oil fields had far exceeded their quotas. Coal output rose nearly 17 per cent between 1973 and 1974, to 482m. tonnes.

Other indicators of China's evolving mining industry include the changing levels of contribution of mining to the nation's GVIO. Since 1949, the Chinese government has measured the performance of all industry to the nation via the GVIO indicator. To assess how a specific industry—such as the mining industry—compares relative to other heavy and light industries in China, a percentage contribution can be examined.

Table 2.4 shows that the contribution of mining-related activities to GVIO in 1957 was 12.7 per cent or, disaggregated into its components, 8.8 per cent from the metallurgical industry, 3.1 per cent from the coal industry, and 0.8 per cent from the petroleum industry. Much of the contribution from the metallurgical industry emanated from iron and steel output. The level of contribution from mining rose to 15.6 per cent by 1984, with a much greater share of contribution coming from the petroleum industry. It was during the mid-1960s when China's growing petroleum industry contributed more to GVIO than the coal industry, and its relative share has increased ever since.

CENTRALLY PLANNED VERSUS FREE-MARKET DEVELOPMENT

Historically, mining in China has been influenced by a number of socio-economic factors, including the system of central planning embraced by the country. The forthcoming chapter describes in detail the impacts of central planning on Chinese mining development by analysing individual economic plans and their influences on minerals and energy output. In addition to output, China's central planning process has had an impact on levels of investment and financing, pricing, taxation, currency exchange rates, and industry distribution. The decision-making power by state-owned[3] enterprises is also a function of the Chinese socialist economy.

Under central planning and state ownership or control, the development of China's mineral and energy industries has been synchronous with the five-year and annual state plans. With a socialist economic system, the objectives and policies of mineral sector development in China are different from those in countries with market forces. Decisions about the development of resources are interrelated with decisions about the nature of society that is desired, and the country's intended role in the international community. Relatively little has been written about the purposes of mining development in centrally planned economies, as under such conditions the determinants (and even definitions) of output allocation, investment, risk, taxation, price, and trade are substantially different from market economies. In their chapter on objectives of the mining industry in developing countries, Bosson and Varon (1977) noted that the information base about the nature, goals, and performance of mining industries in centrally planned economies is very limited and, as such, difficult to characterize.

In a socialist nation, the dominant means of production are under public ownership, and it is the economic sector under public ownership that leads the rest of the economy. Socialist economic development is realized primarily through the plans worked out by the state planning organs. All economic activities in a country like China are aimed at fulfilling the material and cultural needs of the people as defined by the Communist Party. The objectives of the state take precedent over the objectives of the general population. This philosophy extends to the mining industry, where mineral-related activities are administered and controlled by central government agencies in an attempt to advance national economic growth while improving living standards.

In China, mining is part of the social welfare system, where the maximization of social welfare and development (and not necessarily profits) is promoted by the planning organs. With central planning guiding minerals and energy development rather than market forces and competition, mining development is responsive to government policies which commonly aim at achieving political as well as economic objectives simultaneously. By way of comparison, mining development in Western nations is subject to business cycles which are particularly severe in the mining sector, when there are periods of excess supply and demand. Like the

TABLE 2.5. *Change of shares in coal production from mines of different levels of control* (%)

Year	Centrally managed mines	Local mines[a]		
		Total	Local govt.-owned	Collective-owned and private-owned[b]
1949	72.6	27.4	—	—
1957	72.2	27.8	—	—
1958	58.4	41.6	—	—
1960	60.0	40.0	—	—
1965	70.9	29.1	—	—
1970	64.0	36.0	27.0	9.0
1975	58.0	42.0	—	—
1980	55.5	44.5	—	—
1985	47.0	53.0	21.0	32.0
1987	46.99	53.01	20.01	33.0
1988	44.3	55.7	20.50	35.2
1989	43.5	56.5	20.40	36.1

Notes: — No figures available. Figures are rounded.

[a] The number of small coal-mines grew from 16,000 in 1980 to more than 80,000 in 1988. By year-end 1989, 3.4m. people were engaged in local coal-mining activities in China, compared to 3.5m. in centrally managed coal facilities.

[b] One-third of these small-scale mines have a capacity of 10,000 tonnes per year, 30% have a capacity of 10,000 to 30,000 tonnes per year, and 37% have an annual capacity greater than 30,000 tonnes.

Sources: China Coal Industry Publishing House (1991); and Jia Yunzhen (1988*c*).

former Soviet Union, the Chinese mining industry achieved remarkable growth in output as a result of its system of planning, but suffered in other areas such as production efficiency, safety, environmental conservation, and mechanization because of an inefficient planning process attempting to accomplish too many goals with too little information.

To achieve the objectives of social welfare maximization and economic development, China's state-owned enterprises at the central government level exploit the nation's largest and most valuable deposits. In the case of coal, centrally managed enterprises control and operate the nation's largest mines, numbering around 600 and considered the key pillars of the Chinese coal industry. Local coalmines, classified as either local-government-owned, collective-owned, or privately owned, account for just more than half of coal output in China, whereas state-controlled mines produce just less than half (Table 2.5).

The percentage share of China's national coal output from state-owned enterprises has gradually decreased since 1949, demonstrating a gradual shift towards small-scale mining and the effects of decentralization. Government policies in China have traditionally stressed the development of both large and small-scale industries, though local, small-scale operations supposedly were not to compete for resources that were needed by state-controlled enterprises.[4] Strategies for industrial development have clearly affected the quantities of coal produced, their qualities, and distribution of supplies.

Characteristics of China's mining industry that have been influenced by the system of planning and state ownership which predominated until market-oriented reforms began in late 1978 include:

1. *Output allocation* Energy and minerals in China continue to be largely allocated administratively, with market forces playing a minor (but increasing) role in distribution. The primary mechanism for setting allocations of commercial energy has been the materials balance tables associated with the annual plan.
2. *Prices* Prices have been determined by administrative decisions rather than through market mechanisms. They have also been influenced by concerns of income distribution and the need to generate budgetary revenues.
3. *Exchange rates* The value of the renminbi yuan (RMB) versus other currencies is artificially set by the People's Bank of China. The RMB does not float against other currencies and, therefore, does not reflect real value in the global market.
4. *Regional distribution* Political factors in developing a mineral industry are evident when attempts are made to overcome economic inequalities from region to region, even at a higher cost. This is in marked contrast to mining developments in Western nations, where decisions of industry location are based on expected costs and profits.
5. *Taxation* China's income distribution system reflects the tenets of its socialist origins and is largely concerned with an equitable distribution of rents. Very high tax levels and profit remittances to higher authorities prohibit substantial

profits retentions by enterprises, thus creating little incentive to make additional profits through retrofitting or other means, with little or no funds left over to reinvest in new technologies and equipment or other projects at the discretion of the enterprise management.

6. *Financing* Decisions of funding by state agencies are influenced by political motives in addition to economic reasoning. With economic reform, China's government had reduced enterprise dependency on state-allocated funds and implemented a shift to direct bank loans.

The foregoing characteristics must be considered when examining and evaluating the historical development of China's mining industry. The impacts of central planning are indeed far-reaching. Without a market-based economy and pricing system, output decisions do not reflect scarcity of resources, efficiencies in production, and costs of materials and wages. China's economy suffers from an imbalance in resource allocation resulting from the state-controlled pricing system. Exchange rates of the Chinese yuan against other currencies are arbitrarily controlled and, as such, not properly aligned with domestic and international prices or world money market rates. For foreign firms trading with China, especially those whose contracts are denominated and payable in yuan, having a clear understanding of the behaviour of the RMB against international currencies could be critical when huge sums of money are involved.

One major impact of central planning on China's mining industry involves the location of mining enterprises throughout the country, which even today affects the entire industrial structure of China. During the 1950s, government planners had decided to construct large centrally located mining complexes in areas where resources were plentiful. This policy was subsequently revised in the 1960s, when small-scale industry was promoted by Chairman Mao in combination with large-scale operations. In the 1970s and 1980s, planners implemented an overall policy of shutting down the most inefficient small-scale enterprises.

When operating under a system of central planning, reliable and sufficient information is needed by government officials to establish rational mineral and energy policies. The acquisition of such information is, however, costly and time-consuming. Once obtained by state agencies, the information should be shared with producers and consumers to assist them in making more informed decisions. Planners in China have at various times been provided inaccurate or exaggerated information by industrial producers, particularly during periods when accelerated output growth targets were established in annual plans.

Though mining development in China experienced periodic setbacks, its overall history of growth in output was strong and steady. However, as output grew certain aspects of the industry deteriorated, notably efficiency levels and returns on investment. With regard to China's petroleum sector, for example, the capacity added for each additional unit of capital investment has been declining since 1980. Nevertheless, it is difficult to judge the level of success that central planning and control has had in the mineral and energy industries. The definition and characteristics of success differ from one economic system to another. Clearly, for example, labour utilization is considered extremely critical in China with such a huge population base and labour supply, whereas in a sparsely populated country mechanization may receive high priority. What is recognizable with respect to the Chinese mining industry is that trade-offs have occurred between efficiency, safety, the environment, and competitiveness, on the one hand, and the value of subscribing to a planned economy with objectives prescribed by the state, on the other hand.

In actuality, the most reliable way to test the effectiveness of the state-owned mining industry of China would be to remove the protective barriers and regulations of the domestic industry. That is, if the Chinese mining industry were to be fully deregulated and open to international competition and market forces, the response of the domestic industry would be indicative of its efficiency in operations. If, for instance, mining enterprises began to revamp or upgrade equipment on a wide scale, this would suggest that such enterprises were operating at less than optimum efficiency while under state control.

The Chinese economy is now being gradually exposed to market forces, competition, and other elements of market economies including private enterprise and the concept of profits. When China's senior leader, Deng Xiaoping, came to power in the late 1970s, he directed the nation's economy gradually away from its dependence on central planning. Though this movement continued through much of the 1980s, a clampdown on reform in late 1988 revitalized once again central control and planning. The future development of minerals and energy in China will present a difficult choice for government planners where, on the one hand, they could allow market incentives and foreign assistance to facilitate the growth of the industry, while on the other hand they could strive to maintain some measure of control over the industry, which is the fundamental theme behind central planning and the maintenance of the immense power wielded by central planners.

MINING, AGRICULTURE, AND OTHER SECTORS

The economic development of China since 1949 has been fuelled by increases in minerals and energy output. Energy development has taken on particular importance since the early 1960s, when production began at Daqing. Remarkable progress has been made since that time in energy output; lesser gains have been realized in efficiency. Despite the growth in the petroleum sector, however, coal remains the leading energy source in China, accounting for more than three-quarters of the nation's primary energy consumption.

The importance of mining to China's economic development extends far beyond the production of raw materials and products required by industry. Mining and related activities have been integral in China's government policies to stimulate growth of other spheres of the national economy. For example, the iron and steel and the coal industries were part of the 'five small industries' that were developed actively since the mid-1960s as part of China's overall efforts to facilitate rural and local industrialization.[5] Although the focus of China's industrialization policy has been the growth of large-scale industry, an extensive network of rural and local small-scale industries exist that are controlled by local authorities.

The development of China's rural and local industries has been uneven since 1949, owing to changing political motives through the years. Despite uneven development, however, the evolution of small-scale industry has had the major impact of distributing China's industrial base more widely. Production of iron and steel and of coal at the local level has been crucial to the development of Chinese small-scale industry, as well as the regional distribution of industry. Because of the Great Leap Forward initiative, it is estimated that nearly 2 million mud brick blast furnaces (the majority less than 10 feet tall) were constructed by rural workers in any areas available (Tregear 1970). These 'backyard furnaces' led to the production of about 3m. tonnes of low-quality metal in 1958 and 1959, most of which was useless except for the manufacture of simple agricultural tools. Other minerals which contributed significantly to the development of small-scale industry have included nitrogenous and phosphatic fertilizers and cement. These two commodities have been of particular importance to agriculture.

As crucial as the mineral and energy industries have been to industrial development in China, other sectors have also been vitally important to the nation's economy. Of greatest importance to China's overall growth as a nation since 1949 is the agricultural sector, which eventually served as a foundation of the country's economy. The proportion of the agricultural work-force in China's total work-force is nearly 60 per cent today, or greater than 323 million labourers. The share of gross output value of agriculture in the total gross output value of industry and agriculture is just more than 24 per cent. China's current economic reform programme was begun in the agricultural sector, the primary focus of concern at the Third Plenum in December 1978.[6] The nation's rural economy flourished after 1978, with agricultural production increasing in real terms at an average annual rate of 6.6 per cent from 1978 to 1986 (Dernberger 1989). Rapid overall growth in the agricultural sector in the 1980s led to substantial reductions in imports of agricultural products and a richer variety of foods made available to the Chinese population. The success of agricultural reform led to the introduction of economic reforms on a wider basis.

This section briefly highlights some of the many interrelationships and interactions between agriculture and mining. Importantly, mining produces the raw material and energy inputs required to manufacture chemical fertilizers and agricultural machinery. And as the input–output evaluation of Chapter 6 will illustrate, there is a direct linkage between agriculture and mining in China which is particularly evident today, namely the peasant (farmer) miner. As mining output rises in China, farmer income also tends to increase. With the decentralization mode of much of the 1980s, farmers were allowed to engage in mining activities to supplement their incomes. This had an impact on China's rural labour force development and movements, as there are today an estimated 122,500 small mines operated by farmers in China, with an output value of approximately one-sixth of the nation's total. Peasant farmers have typically mined coal, gold, copper, lead, and zinc, among a host of other metals. Mining is a profitable venture for farmers, and indeed China's grain harvests since 1984 have suffered owing to the growing dissatisfaction of peasants in cultivating grain crops, inasmuch as other branches of farming or mining activities bring them comparatively higher profits.

1. *Overview of China's Agricultural Development*

Figure 2.3 illustrates the changing structure of China's economy by sector since 1952 based on percentage composition of National Income defined as Net Material Product. Figures for 1952–65 and 1966–78 represent percentages of total accumulated National Income for the period indicated. As depicted, the contribution of the agricultural sector to China's National Income dropped during the entire period represented, whereas industry grew strongly until 1979, after which its contribution declined slightly. Between 1952 and 1978, investment in construction projects increased from 26.5 per cent to 40.7 per cent of the state's expenditures, and the net value of fixed assets in state-owned industry rose an average 13.9

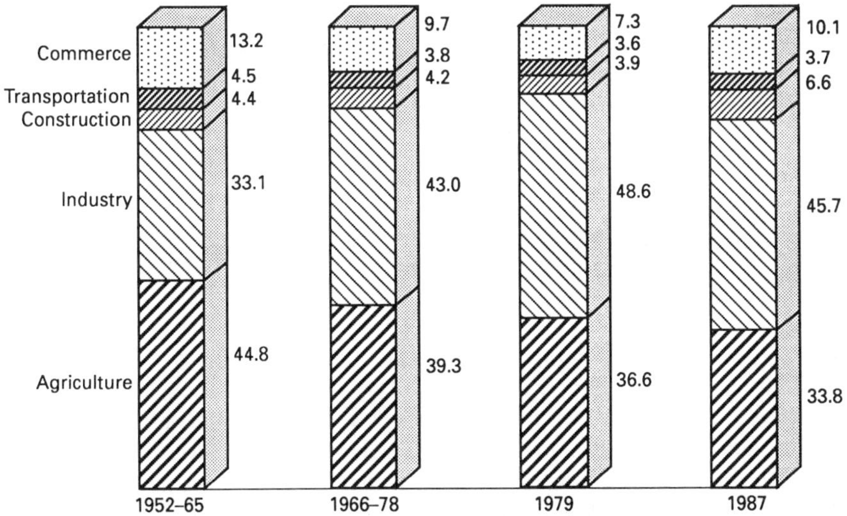

FIG. 2.3. Changes in the relative structure of China's economy as based on composition of National Income, 1952–1987

per cent annually (Dernberger 1989). During the period, China's National Income growth averaged 6 per cent per year, and this was largely attributed to the growth of industry which averaged 11.3 per cent annually, as compared to the 2.2 per cent growth rate for agriculture. In contrast, farmer incomes were not increasing at a rate anywhere near the growth rate for the overall economy. Although prices for farm products generally rose between 1952 and 1978, gains in peasant income were small, particularly after 1965. In agriculture, the increase in purchased inputs per unit of output implied that peasant net incomes were not rising even though gross outputs were increasing. One estimate places the gains in income derived from collective sources by peasants at half of 1 per cent per year between 1965 and 1977 (Lardy 1986). Slow growth in farm output was accompanied by extremely weak growth of peasant income.

Preceding the initiation of economic reform in 1979, China's agriculture had been dominated by state planning and the commune system. Agricultural output grew slightly during the years 1952 to 1978, whereas productivity declined. As China's agricultural sector expanded in output terms, it also absorbed large numbers of new workers owing to collectivization[7] and the introduction of industrial inputs into farm production activities (Rawski 1979). Until 1956, family farming remained as the dominant type of agricultural organization in China. Two years later, as part of the Great Leap Forward initiative, the rural labour force was organized into large-scale collectives or people's communes. Owing to mismanagement, lack of incentive, poor weather, and other factors, the early harvests of the people's communes suffered tremendously, prompting decentralization in the early 1960s, which returned authority over agricultural production and income distribution to smaller units roughly the size of the co-operatives before 1956.[8]

Except for the periods of extreme economic turmoil, industry in China grew strongly during much of the 1950s, 1960s, and 1970s. From the late 1950s increasing quantities of industrial inputs (including power, machinery, building materials, steel, petroleum products, and chemical fertilizer) were being introduced into the agricultural sector. Between 1957 and 1978 the growth in farm machinery was astounding, although beginning from a very small base (Table 2.6). The growth of three major types of farm equipment—tractors, power tillers, and irrigation and drainage equipment—grew at an average annual rate of 24 per cent between 1957 and 1978. In terms of millions of horsepower, the growth in these three types of farm equipment rose from an estimated 1m. hp in 1957 to nearly 93m. hp in 1978. By 1978 the farm machinery industry in China had grown to considerable scale,[9] where its output value was more than 160 times that in 1952. The nation manufactured more than 110,000 tractors in 1978, compared to fewer than 10,000 in 1965.

Although in physical terms and horsepower the growth of industrial inputs into agriculture was dramatic during the 1950s, 1960s, and 1970s, such inputs did not necessarily lead to high levels of agricultural mechanization. The proportion of total steel production devoted to agriculture, farm machinery, and farm machinery maintenance never exceeded 14.1 per cent in any given year between 1953 and 1979 (Liang 1982). Steel and steel products used in the construction of farm machinery were sometimes of such low quality that the equipment could not be operated.

The increased usage of machinery released substantial numbers of rural labourers from their tasks; many of these workers were absorbed into other farm or agriculture activities (reducing the marginal product of labour but increasing its total product). Table 2.7 contains labour force figures for agriculture for the 1952 to 1979 period. Although the percentage contribution of agricultural labour to China's total labour-force dropped over time, millions of workers were absorbed into agriculture simultaneously. Estimates are that more than 100 million workers were absorbed into agricultural activities between 1952 and 1979 (Rawski 1979; Riskin 1987), amounting to an increase in the labour-force of more than 40 per cent, yet there were no significant gains in cultivated land area.

Along with the mechanization of agriculture in China came a dramatic growth in the production of chemical fertilizers. Fertilizers are manufactured materials, commonly containing nitrogen, phosphorus, and potassium, applied to soils to boost plant nutrition levels. By-products from the steel industry and wastes from food-producing operations have traditionally been major fertilizer sources in China, although these have become less important with development of efficient manufacturing processes for commercial fertilizers. In terms of mineral and energy inputs, natural gas is the primary source of hydrogen for ammonia production, which requires about 40,000 cubic feet of gas per ton of ammonia. Natural gas also provides carbon dioxide for urea production, another nitrogen fertilizer.

Phosphate-based fertilizers are manufactured using phosphate rock as the major raw material input. Because of the insoluble nature of the phosphate mineral fluorapatatite, chemical treatment with strong acids (sulphuric, phosphoric, or nitric) is necessary to produce soluble phosphate products. In contrast, potassium-based fertilizers require potash ore, which can be used directly as potassium chloride (often called muriates) or sulphate without extensive chemical conversion.

Output of chemical fertilizers in China rose from 200,000 tonnes in 1953 to over 40m. tonnes in 1978. The 23.6 per cent average annual growth in chemical fertilizer output during this period elevated China among the world's leading nations of fertilizer production, consumption, and international trade. China's long-term policy of fertilizer production rested upon producing as many varieties as possible. In the 1950s and early 1960s there was a consolidation of some older plants and a purchase of new plants from neighbouring countries. During this time the larger plants generally produced nitrogenous fertilizers, including urea, ammonium sulphate, and ammonium chloride. Substantial growth in the construction of small and medium-sized nitrogenous plants throughout China occurred in the 1960s and 1970s.

TABLE 2.6. *Growth of industrial inputs into China's agricultural economy, 1957–1978*

Year	Rural power consumption (billion kWh) (1)	Output		Inventory data (m. hp)				
		Small-scale cement (m. tonnes) (2)	Chemical fertilizer (m. tonnes)[a] (3)	Irrigation and drainage equipment (4)	Tractors (5)	Power tillers (6)	Total for three categories	
							Million hp (7)	Hp per cultivated ha (8)
1957	0.1	—	0.8	0.6	0.4	—	1.0	0.01
1962	1.6	1.6	2.8	5.8	1.5	—	7.3	0.07
1963	2.1	2.3	3.9	6.4	1.7	—	8.1	0.08
1964	2.5	2.2	5.8	7.3	1.8	—	9.1	0.08
1965	3.2	5.4	7.6	8.4	2.2[b]	—	10.6	0.10
1970	NA	11.5	14.0	16.9	4.5	0.3	21.7	0.20
1971	9.6[c]	12.4	16.8	20.0	5.6[b]	0.4	26.0	0.24
1972	NA	18.3	19.8	24.0	6.6	0.7	31.3	0.29
1973	NA	20.5	24.8	30.0	8.5	1.1	39.6	0.37

1974	NA	21.3	24.9	36.0	10.9	1.6	48.5	0.45
1975	14.4[d]	27.7	28.8	43.0	12.3	2.2	57.5	0.54
1976	NA	29.6	24.3	46.5[b]	14.6	3.2	64.3	0.60
1977	NA	36.0	38.0	60.0[c]	16.0[f]	5.6[b]	81.6	0.76
1978	27.0[g]	NA	48.0[h]	65.6[c]	19.1[e]	8.0[i]	92.7	0.87

Notes: — negligible. NA not available.

[a] Based on product weight, not nutrient weight.
[b] Linear interpolation between figures for consecutive years.
[c] Reported to be six times the figure for 1962; cited in Wiens (1980: table 17).
[d] Based on increase for 1965–75 reported in *Peking Review* (8 Oct. 1976).
[e] *Beijing Review* (1979*a*, *b*). Tractor hp is assumed to have grown in proportion to the 19.3% increase in the number of tractors reported for 1977/8.
[f] Previous year's inventory less 5% for retirements (assumed) plus 1977 output, as estimated in US National Foreign Assessment Center (1977: 20).
[g] Calculated from Chen Nai-ruenn (1967: 186) and *China Reconstructs* (1979).
[h] *Beijing Review* (1979*a*, *b*).
[i] Bureau of Foreign Affairs (1978: 4) reports inventories of 800,000 power tillers. The average size of power tillers is assumed to be 10 hp.

Sources except as noted: Column 1, Kang Chao, 1970; Columns 2 and 3, US National Foreign Assessment Center (1977: 13); Column 8, based on a constant total of 107m. cultivated ha cited in Perkins (1975). Compiled by Rawski (1979: 80–1).

TABLE 2.7. *Changes in sectoral composition of output and labour-force, 1952–1979*

	1952		1957		1965		1975		1979	
	m.	%	m.	%	m.	%	m.	%	m.	%
Labour-force	207.3	100.0	237.7	100.0	286.7	100.0	381.7	100.0	405.8	100.0
Agriculture	173.2	83.6	193.1	81.2	234.0	81.6	294.6	77.2	299.3	73.8
Non-agriculture	34.1	16.4	44.6	18.8	52.7	18.4	87.1	22.8	106.5	26.2
	bn. Y	%	bn. Y	%	bn. Y	%	bn. Y	%	bn. Y	%
GVAIO (bn. renminbi)	81.0	100.0	124.1	100.0	223.5	100.0	446.7	100.0	637.9	100.0
GVAO	46.1	56.9	53.7	43.3	83.3	37.3	134.3	30.0	189.6	29.7
GVIO	34.9	43.1	70.4	56.7	140.2	62.7	312.4	70.0	448.3	70.3
National Income (current prices)	58.9	100.0	90.8	100.0	138.7	100.0	250.5	100.0	335.0	100.0
Agriculture	34.0	57.7	42.5	46.8	64.1	46.2		39.0	131.8	39.3
Industry	11.5	19.5	25.7	28.3	50.5	36.4		45.0	153.6	45.9
Construction	2.1	3.6	4.5	5.0	5.3	3.8		4.0	13.0	3.9
Transport	2.5	4.3	3.9	4.3	5.8	4.2		4.0	12.1	3.6
Commerce	8.8	14.9	14.2	15.6	13.0	9.4		8.0	24.5	7.3

Notes: The agricultural labour-force figure for 1975 is estimated by multiplying the rural commune labour-force by 0.984, which is the average for 1970 and 1979 of the proportion of commune labour engaged in agriculture.

GVAIO, GVAO, GVIO, and National Income figures obtained from Table 1.1.

Sources: Riskin (1987: 270); Emerson (1982: 224–67).

In concurrence with the building-up of a manufactured fertilizer industry in China, other industries were spawned including those of petrochemicals and mixed fertilizers. Petrochemicals include those chemicals derived from petroleum which serve as feedstocks for the manufacture of a variety of plastics and synthetic rubber. China's petrochemical industry grew during the 1960s and 1970s and helped to provide for the packaging, storage, and transport of chemical fertilizer materials. In the 1960s China's leadership made an effort to take advantage of the country's abundant petroleum and petroleum-derived raw materials by building petrochemical and fertilizer plants. Japanese and Western equipment, and in some cases complete plants, were imported into China for the expansion of its petrochemical industry.

2. Balancing Development of Agriculture and Industry

China's leaders have always considered the interaction of mining and minerals with agriculture to have an importance far beyond the usage of steel in the production of farm machinery or the development of a fertilizer industry. Domestic industry had to produce the raw materials and other inputs needed to transform agriculture over time because of the low reliance on foreign trade. Co-ordinating the development of agriculture and industry was deemed necessary in China because combined they are the materials production backbone of China's economic structure. As agriculture expanded and developed, its linkage to industry also grew. Agriculture and industry became interdependent, with each serving as both a constraint and a stimulus on the other. Chinese leaders therefore believed that establishing a balanced relationship between the two was needed to provide for a balanced development of the whole economy.

In China industrialization would have been difficult without the modernization of agriculture and, simultaneously, without a modernized industry, agricultural development would have been difficult (Liang 1982). Since the 1950s agriculture served not only as a provider of some raw materials, capital funds, and labour for industrial expansion, but it was also a growing market for industrial inputs. Heavy industry, including the minerals-related sectors, supplied China's agriculture with raw materials for machinery, chemical fertilizers, and fuel. With regard to the latter, rural electrification in China spread rapidly in China's rural areas in the late 1960s and 1970s, contributing to a large increase in rural power consumption. In terms of kilowatt-hours, China's rural power consumption grew nearly sixfold between 1962 and 1971 (Wiens 1980).

At the time of formulating China's First Five-Year Plan, the industrial leaders felt that the industrial and agricultural bases of China were relatively weak and small-scale (Liang 1982). With many options before them, they knew that while they gave priority to the development of heavy industry, efforts had to be made to develop agriculture. Table 1.1 of the introductory chapter briefly outlined the priorities of China's many five-year plans in terms of heavy or light industry, or agriculture. The priorities adopted during each five-year plan affected the balance of development between industry and agriculture. During the Great Leap Forward, as an example, emphasis on the rapid development of heavy industry (particularly iron and steel) meant that other sectors suffered. The demand that steel output double from 5.35m. tonnes to 10.7m. tonnes in 1958 required large amounts of capital and labour at the expense of agriculture and light industry. From 1958 until economic readjustment in the mid-1960s, agricultural production dropped significantly. This occurred again periodically during the Third and Fourth Five-Year Plan periods when the concentration of steel production was once again reasserted.

Table 2.8 shows the relative distribution of capital construction funds to agriculture, light and heavy industry, and metallurgy between 1953 and 1991. China's historical emphasis on stimulating rapid growth of heavy industry, and particularly iron and steel, is obvious in the table. The proportion of funds devoted to agriculture increased to a high during the 1963–5 Period of Readjustment, but generally remained at lower than needed levels throughout the period. The distribution of funds favoured agriculture over metallurgy during the period of economic readjustment, and the Third and Fifth Five-Year Plan periods.

TABLE 2.8. *Relative distribution of capital construction investment of state-owned units in agriculture, light industry, and heavy industry, 1953–1991* (%)

Years (Period)	Agriculture	Light industry	Heavy industry (incl. metallurgy)	Metallurgy[a]
1953–7 (First)	7.1	6.4	36.1	7.9
1958–62 (Second)	11.3	6.4	4.0	4.0
1963–5 (Adjustment)	17.6	3.9	45.9	8.0
1966–70 (Third)	10.7	4.4	51.1	10.1
1971–5 (Fourth)	9.8	5.8	49.6	9.8
1976–80 (Fifth)	10.5	6.7	45.9	8.1
1981–5 (Sixth)	5.0	6.9	38.5	6.2
1986–90 (Seventh)	3.3	7.5	44.3	5.5
of which:				
1986	3.0	7.0	38.2	4.6
1987	3.1	7.4	43.5	5.9
1988	3.0	7.8	43.8	6.1
1989	3.3	7.9	45.1	5.8
1990	3.9	7.1	48.8	5.0
1991 (Eighth)	4.0	7.2	47.0	5.9

[a] Proportion of investment in all sectors of the national economy, not only heavy industry.

Source: State Statistical Bureau (1981–92 (1986–92), Beijing and Hong Kong).

3. *Longer-Term Agricultural Reform*

Based on actual levels of capital construction funds, the share of accumulated investment over the period 1953 to 1979 in heavy industry was about 50 per cent, with the metallurgical industry receiving about 12 per cent. Some Chinese economists (Liang 1982) feel that the emphasis of capital construction funding on heavy industry and iron and steel did not stimulate the anticipated growth in other spheres of the economy, and instead resulted in imbalance between industry and agriculture, and an irrational economic structure. Perhaps this sentiment is one reason why in 1979 many Chinese leaders stressed the need for China to readjust heavy industry to orient itself more towards agriculture and light industry. Calls were made for heavy industry to produce higher-quality, more efficient farm machinery and chemical fertilizers, as well as provide China's countryside with more energy. An ultimate goal is to increase agricultural productivity while attempting to keep China's rural labourers employed.

Specifically, the economic policies put forward at the Third Plenum in China largely centred around agriculture reform. Agricultural growth and diversification since the late 1970s has been an impetus for overall economic growth. China's rural reforms implemented after 1979 can be disaggregated into four parts (Riskin 1987): (1) significant price increases for farm products; (2) greater autonomy of decision-making authority for the collective; (3) the abandonment of a strategy of forced local self-sufficiency in grain with one of encouraging diversification and specialization; and (4) rapid decollectivization of labour and income distribution. As a result of these changes, China's rural economy significantly expanded from 1978, and agricultural gross output values (GVAO) grew at nearly unprecedented levels to 1986.

Although since 1986 the rate of agricultural expansion in China has slowed and the surplus in agricultural trade has fallen, the overall growth in agriculture since reform was implemented has led to a drop in imports of agricultural products and an elimination of the rationing of many basic foodstuffs to the urban population. In terms of industry response, the success of the agricultural reforms initiated in 1979 importantly served as a stimulus to introduce economic reforms on a wider scale. Industrial sector reforms were thus begun at a later date. Although the initial gains were not as impressive as those in agriculture, industrial output, including minerals and energy, grew strongly in the 1980s.

Some of the linkages in China between agriculture and industry, and particularly mining, have been highlighted in this section, including enhanced raw material inputs for mechanization, labour movements, and changes in direction of capital construction funds. There are, in addition, linkages within each individual sector which are important. In the past, when the emphasis was placed on heavy industry in an economic development plan, China's agriculture and light industry tended to suffer. As a system of input and output flows, when one sector of China's economy expands (or contracts), other sectors are affected. In the years ahead, efforts will be made to develop agriculture and industry in as co-ordinated a manner as possible. China's agricultural policy will emphasize the strengthening of agriculture as the foundation of China's economy, while industrial strategies will stress the restructuring of industry in order to stimulate agricultural growth. Clearly, the process of industrialization in China is facilitated as well as constrained by developments in the nation's agricultural sector.

SUMMARY OF DEVELOPMENT HISTORY

This chapter has presented a discussion of the historical development of the Chinese mining industry and its impact on the nation's overall economic well-being. Mining and metallurgy have undergone substantial growth in China since 1949, resulting from a pronounced commitment by the government to expand the industry, a huge resource inventory, a need to supply raw material inputs to a large industrial base, and a desire to maintain a high level of self-sufficiency in resources. As early as 40 years ago the steel industry of China was relatively small and lacking efficient and operable machinery, the coal industry was just beginning to expand but accounted for only 2 per cent of world output, and the country was trying to rehabilitate itself after the war-torn years of the 1930s. Today, the mining industry has evolved to become the fourth largest throughout the globe, helping to fuel the growth of one of the fastest-growing economies in Asia and the world.

During the past four decades the Chinese government was determined to develop and expand the metallurgical, coal, and petroleum industries, as they received about 34 per cent of total capital investment into industry. Investment levels varied through time, depending on the different emphases of the five-year economic plans. Development of China's rich minerals base has always been considered a top priority among government officials, and vital for furthering the growth of agriculture and industry.

The minerals production history of China was depicted in Figure 2.2 by way of indices of the physical volume of minerals and energy output. Though generally positive, mining development was periodically affected by setbacks owing to irrational industrial policies, political instability, or economic downturns. As mining output rose, problems persisted in China's mining industry, including low recovery rates in mining and processing, low efficiencies in materials utilization, environmental degradation, and declining conditions of safety.

The decade of the 1960s was particularly volatile for mining, as output reached a peak in 1960, only to be followed by a significant slowing in nearly all industrial activities resulting from the Great Leap Forward. After the industry recovered, the Cultural Revolution commenced which again introduced instability and chaos into minerals exploration and development activities. In total there have been four periods in the historical development of the Chinese mining industry, with each period characterized by consistent rates of change in output, and specific mineral commodities and industrial policies.

As China's mining industry evolved, its impacts on the nation's overall economy changed accordingly. The metallurgical, coal, and petroleum industries accounted for 18.2 per cent of China's GVIO in 1990, compared to just more than 12 per cent in the 1950s. As mining progressed, its importance to other sectors such as agriculture became more pronounced, and the distribution of large and small-scale industry was increasingly affected. When the allocation of capital construction funds to heavy industry (including metallurgy) expanded, less funds became available for agriculture and light industry. Economic and industrial policies helped to guide the minerals development history of China, primarily with an aim of fostering economic growth. The impacts of the individual five-year plans on the mining industry are described in the next chapter.

NOTES

1. In a lengthy study of the Chinese mining industry conducted by the Foreign Minerals Division of the US Bureau of Mines (1948), it was estimated that coal was the most valuable mineral or energy commodity produced in 1933, with an estimated output value of 4.5 times the second most valuable commodity—salt. Readers are referred to this study and Dorian (1987) for further discussion on historical output values.
2. Throughout this book dollar values are to be understood as US dollars. An independent study conducted by the author (Dorian 1987) estimated that China's minerals and energy output in 1983 was worth an estimated $51.0bn., which closely resembles the figure published by Callot (1985) and thus provides some level of confidence in its accuracy. In the 1987 study, output levels of 49 mineral and energy commodities produced in 1983 were valued using Chinese prices where available and major world producer prices where unavailable. The entire study calculated values of metals, non-metals, and mineral fuels production in China between 1949 and 1985.
3. Throughout this book, the use of the term 'state-owned' refers to an enterprise owned by the Chinese state and an organ of the state at any government level, including central, provincial, county, prefecture, or city. 'Centrally owned' would imply the enterprise is owned by China's central government in Beijing or one of its ministries. 'State-run' or 'state-controlled' indicates that the enterprise is operated by the state at any one of several government levels. State ownership dominated China in the 1950s, 1960s, and 1970s; by the 1980s, however, collective-owned and privately owned enterprises began to flourish in the nation.
4. 'Walking on two legs', or *Lian Diao Tai Zhou Luo*, has been a fundamental policy to simultaneously develop both large and small industries, in order to enhance localized

economic strength, reduce transportation requirements, and enable legislation to better serve agriculture (K. P. Wang 1975). With minerals, this policy implied trying to utilize resources in an optimum manner by simultaneously developing large and small facilities.

5. For a comprehensive discussion of China's system of small-scale industry, readers are encouraged to read *Rural Small-Scale Industry in the People's Republic of China*, by the American Rural Small-Scale Industry Delegation (Perkins *et al.* 1977).
6. Much has been written on China's agricultural development (see e.g. Zhu Xigang and Chapman 1988; Lardy 1986; Rawski 1979; Perkins *et al.* 1977; Kuo 1976; and Chu Li and Tien Chieh-yun 1974), and readers are referred to these publications for detailed discussion.
7. For one of many detailed accounts of collectivization, see Donnithorne (1967).
8. Numbers of units are reported in a 'Briefing on China's Agriculture', distributed by the Bureau of Foreign Affairs, Ministry of Agriculture and Forestry (1978: 2). Estimated unit sizes are based on a rural population of 789m. derived by projecting the estimated 1975 rural population forward at an annual rate of 2% and on rural household size of 4.4 persons. These projections were made in Crook (1975).
9. In 1978 China's tractor density was an estimated 5.6 units per thousand hectares, compared to a median for all developing countries of 2.6 units that same year (International Agricultural Development Service 1981).

3

Impact of Central Planning on Historical Mining Development

SYSTEM OF CENTRAL PLANNING

Minerals and energy development in a country is strongly influenced by the nation's economic and political system. In the United States and other Western nations, such development is ordinarily guided by free-market conditions which are established by supply and demand forces. Such forces are, in turn, expressed in terms of mineral and energy prices. When the price of a commodity is high, private sector companies tend to produce more in order to reap larger profits. When a large number of private firms enter a market and produce more, the price of that good will eventually stabilize and then decrease, until an equilibrium between supply and demand has emerged.

Unlike a market-based economy, China's socialist economic system up until the beginning of market-oriented reforms in the late 1970s was not necessarily guided by prices, or, in other terms, by the direct interaction of supply and demand forces. Mineral and energy resources in China remain exclusively under the domain of the 'whole people' as represented by the state in the form of either central or local governments, and the state largely directs and oversees the exploration and exploitation of resources in the nation, and prohibits any organization or individual from encroaching upon or destroying mineral or energy resources by any means. With regard to the prospecting for and exploitation of mineral and energy resources, the state implements the policy of centralized planning, rational overall arrangement, comprehensive prospecting, reasonable exploitation, and multi-purpose utilization of mineral resources. In short, the state influences and controls the exploitation of mineral and energy resources in China under the overall guidance of the Chinese Communist Party (CCP). The state also controls directly the allocation of output and services of the mineral and energy sectors to ensure that adequate supplies are provided to preferential users. Under such a system, prices perform a largely accounting function and have little or no allocative role.

China's socialist economy is stated by the post-Mao leadership to be a planned commodity economy, and the national economy is supposed to be managed and administered according to objective rules to invigorate the economy for a maximum return (Li Honglin 1986). But the extent to which rational or objective economic criteria in economic decision-making were adhered to has varied widely over the past 40 years. Since the launching of the Great Leap Forward in 1958 and until the Third Plenum of the 11th Central Committee in December

1978, the dominant leftist ideology often subordinated rational economic principles to political objectives and considerations. Development policy, for example, was for a long period aimed at establishing self-sufficient and basically autarkic units at all levels and regions of the economy (the 'large yet comprehensive, small yet comprehensive' policy). Such a policy militated against the principle of development according to a region's comparative advantage and consequently resulted in significant economic losses and a distorted sectoral and regional pattern of development.

In this context, minerals and energy planning in China after 1949 was typically part of the nation's five-year plans, as well as specific economic reforms and annual economic plans. Figure 2.2 in the last chapter provided some indication that central planning (and the lack of orderly planning) strongly affected mining output in China throughout the 39-year period 1952–91. As was shown, there are distinct output trends, both positive and negative, of major commodities throughout the history of China's mining development. These trends were disaggregated into four major periods, with three of them being characterized by positive, steady output growth, and one—Period II—noted for fluctuating production levels. During Period II mining output, barring a few commodities like crude oil, declined briskly in the early 1960s as a result of the chaotic economic planning and devastating impacts of the Great Leap Forward and the People's Commune Movement. Although the Great Leap Forward and People's Commune Movement began in 1958, the disastrous effects of poor management at mines, waste and inefficiencies, and over-use of machinery were not strongly felt throughout China's mining industry until 1960 and thereafter. A similar lag of devastating impacts occurred in respect to the first phase of the Cultural Revolution; these lasted until about 1971. Since that year, China's mining industry has experienced relatively strong and stable growth.

To fully understand the impacts of central planning on minerals and energy development in China, each individual five-year planning period is analysed in this chapter, in addition to other major economic and political movements. Although not yet complete, the Eighth Five-Year Plan is also evaluated. All of these plans and movements are discussed primarily in terms of policies which directly affected China's heavy industry and, more particularly, the mining industry, and the resulting industrial and minerals/energy output during the respective planning periods.

MINERALS AND ENERGY PLANNING PROCESS IN CHINA

In China, the State Planning Commission has the final judgement in setting priorities for minerals and energy exploration and development. These priorities are based on the national economic plans and objectives established by the State Council and National People's Congress—with overall guidance and direction by the CCP. Although the State Planning Commission is responsible for establishing policies for minerals and energy development, it does so in conjunction with the various mineral-related ministries in China and the provincial government leaders. Specialists from the ministries and provincial governments' offices advise and assist the State Council and State Planning Commission in setting mineral and energy strategies for five-year plans.

In order to establish economic and long-term plans, several bureaux exist within the State Planning Commission which focus on particular subject areas and problems. The National Land Bureau is one such department which concentrates its efforts on minerals and energy planning, and helps co-ordinate the strategies of all ministries in China and integrates these strategies into national economic plans. There are eight major ministries or corporations which provide input into national mineral and energy plans, including the Ministry of Geology and Mineral Resources, Ministry of Energy, Ministry of Metallurgical Industry, Ministry of Materials, Ministry of Construction, Ministry of Chemical Industry, CNNC, SINOPEC, plus various other light industries. All of these organizations are engaged in mineral and energy development activities.

Mineral and energy strategies of China are primarily based on (1) domestic requirements for industry, (2) employment needs, and (3) technology availability (particularly mining extraction equipment). Other factors considered include export opportunities to generate foreign exchange, resource availability or endowment, and infrastructural problems. The minerals and energy planning process in China is very complex, involving a multitude of organizations at several levels of authority. The entire process is depicted in Figure 3.1.

As shown, three basic organizational tracks exist where mineral and energy plans are formulated and subsequently implemented: ministry–enterprise, State Planning Commission–planning section/division, and provincial government–enterprise. Thus planning in China occurs along a dual-track system: sectoral planning by the ministries (called in Chinese *tiaotiao* or line planning, and territorial planning by the local governments (called *kuaikuai* or bloc planning). Sectoral plans of the ministries and territorial plans of local governments are then co-ordinated and integrated by the State Planning Commission. Large enterprises or corporations (such as SINOPEC) also engage in planning and implementation, particularly the ministerial-level organizations enjoying significant political and economic decision-making power as well as funding. Adjacent to these lines of planning is the CCP, which plays a supervisory role at all levels throughout the entire planning process.

In general, a plan is developed at the highest level of authority based on inputs and suggestions from organizations lower in stature in the planning hierarchy. The extent of authority declines from the top in Figure 3.1, and during a planning process policy directives move down with the direction of authority. In response to a directive, proposals, ideas, and statistics move upward. Once a plan is developed, it is carried through at each level of authority, though the motives and objectives at the different levels are sometimes different. Chapter 7 will describe

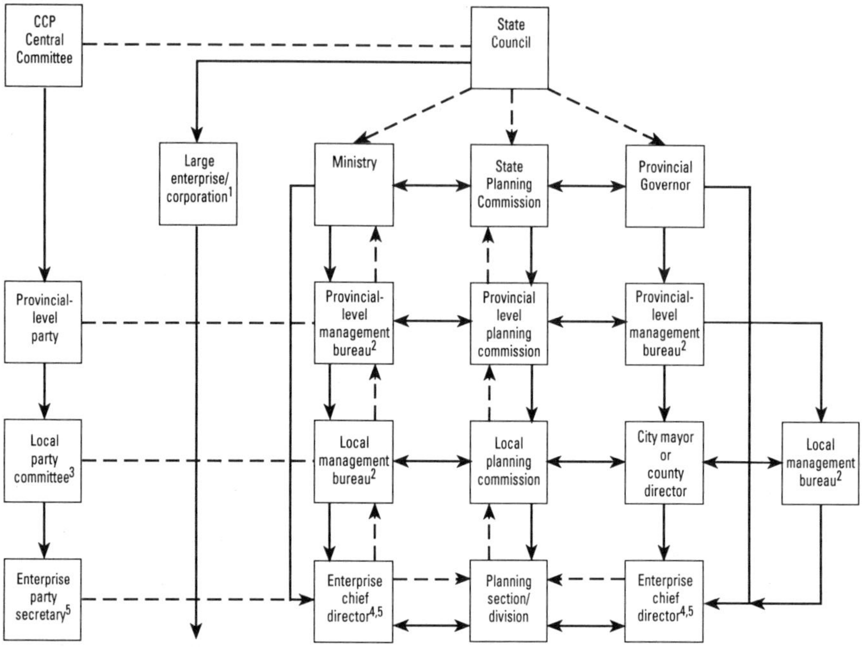

– – Denotes supervisory role of Party

↔ Implies lateral communication occurs between organizations

↓ Implies policy directives move downward

⇡ Implies proposals/ideas move upward at the initiative of the enterprise

Notes: Extent of authority declines downward in diagram.

The State Council only becomes involved if a problem among the ministries, the State Planning Commission, or provincial governors cannot be resolved.

During a planning process, directives move downward with the direction of authority, and suggested proposals, ideas, and statistics move upward in response.

Suggested proposals, ideas and statistics also move upward on the initiative of local enterprises. The initiatives, in addition to being routed along the ministerial track, are also routed along the planning commission track for co-ordination and approval.

Ministries commonly co-ordinate policies directly with major enterprises whereas smaller enterprises, belonging to a particular ministry, may remain under control at the provincial level.

[1]Large enterprise/corporation can be of ministry level, such as the China National Petrochemical Corporation (SINOPEC), or below.

[2]Same organization.

[3]Local implies city (*shi*) or county (*xian*) or, in some cases, prefectures (*qu*).

[4]Enterprise chief director may report to a ministry or provincial governor and not necessarily to both.

FIG. 3.1. Minerals and energy planning process in China

some of the different motives between central ministerial-level and provincial/local-level enterprises.

With the ministry–enterprise line of planning, a specific mineral-related ministry will forward its recommendations and/or obligatory targets to provincial-level industrial management bureaux. The ministry will send two sets of guidance/obligatory targets to its lower levels: one set directly to more important enterprises under its control, another to the industrial bureau of the provinces for co-ordination purposes so that activities for ministry-controlled enterprises may be integrated or co-ordinated with activities under provincial control. The direction of authority and decision-making power moves and declines respectively from a ministry to a provincial-level management bureau, local management bureau, and specific chief director. The input received from a mine enterprise eventually works its way back through the provincial-level management bureau to a specific ministry.

In cases of conflicts between ministerial and provincial plans, either one or the other takes precedence, depending on the historical period. In 1953–7, 1961–6, 1973–8, sectoral (ministerial) plans enjoyed precedence; but in other periods (for example, the decentralizations of 1958–60, of 1970–2 and since 1978) territorial plans of provinces enjoyed priority (Zhou Taihe 1984).

A similar procedure of planning occurs with the State Planning Commission; guidance and obligatory targets are sent from the State Planning Commission to provincial level and local planning commissions, which subsequently send their comments back to the State Planning Commission for review. The State Planning Commission will send its directives to ministries and local governments, and the ministries and local governments will then disaggregate draft targets for enterprises under their respective control. After a process of iteration and negotiation, revised draft plans are returned upwards by ministries and local governments to the State Planning Commission.

Once the plans have been reviewed at the provincial and local levels along all of the planning tracks, the ministries and provincial governments discuss their agendas directly with the State Planning Commission. Upon approval by the State Council of the comprehensive plans prepared by the State Planning Commission, final plans are returned to ministries and local governments for transmission down respective networks for implementation. It is between these three bodies of government, plus large enterprises/corporations, that mineral and energy plans and objectives are finalized. If a specific ministry or provincial governor cannot agree on a strategy with the State Planning Commission, the State Council may intervene to settle this issue. Otherwise, the State Council does not become involved in the negotiations of mineral and energy policies.

To comprehend the impact and effectiveness of China's socialist economic planning of minerals and energy development, the following discussion reviews historical plans for economic and industrial development in China, for the purpose of emphasizing their influences on the mineral and energy industries. The discussion provides only a brief review of China's development plans and strategies, including the Period of Rehabilitation through the Eighth Five-Year Plan. For more in-depth analysis of China's historical plans, readers are referred to Riskin (1987), Liu Suinian and Wu Qungan (1986), Su Kaiming (1986), Xinhua News Agency News Bulletin (1986), Foreign Language Press (1984), and the State Statistical Bureau (1985; 1960).

As some of the five-year plans of China were never fully published, notably the Third, Fourth, and Fifth Five-Year Plans, discussion on the impacts of these plans on minerals and energy development is limited.

THE PERIOD OF REHABILITATION (1949–1952)

In 1949 the Chinese government decided that the following three-year period would be years of gradual rehabilitation, rebuilding, and improvement. The government realized that it had a major task to embark upon in reconstruction of the economy (restoring output to pre-war years), improving productive services, revamping major transportation systems, and rejuvenating administration operations. The newly established and inexperienced government also attempted to re-establish heavy industry in China, particularly in three north-eastern provinces of China—Liaoning, Jilin, and Heilongjiang. However, this attempt was hampered by increasing demands for military supplies because of the unexpected Korean War begun June 1950 (Perkins 1966).

The Chinese Civil War (1946–9) had a damaging effect on industrial production in the nation, particularly in 1949, the pinnacle of the war. During that year Chinese industrial production fell to its lowest level since 1945. Because of this dilemma, the Chinese government enacted a liberal policy towards the industrial sector by providing incentives to workers and owners. Several private enterprises were allowed to continue during the years of rehabilitation, even though the Chinese government had wanted to implement public ownership of production services.

The policy at this time was 'to use, restrict and gradually transform the private (capitalist) sector' so that it could play a role in economic recovery. In June 1950, China's government at various levels began to readjust industry and commerce in accordance to the directives of Mao Zedong (Hsueh Mu-chiao, Su Hsing, and Lin Tse-li 1960). This included readjusting the relationships between the different sectors. The state sector controlled the capitalist sector through the process of exchange, incorporating its production into the overall state plan. Through 1952, the primary objective of Chinese Communist financial policy was to gain fundamental control over the nation's finances so as to form a base from which to establish more complete control over all areas of the economy and to stabilize prices (Perkins 1966).

TABLE 3.1. *Estimated production of selected industrial items during the Period of Rehabilitation (1949–1952)*

	1949	1950	1951	1952	% increase 1949–52
Cement (tonnes '000)	660	1,410	2,490	2,860	333.3
Chemical fertilizers[a] (tonnes '000)	6	15	28	39	550.0
Coal (m. tonnes)	32	43	53	66	106.3
Crude oil (tonnes '000)	120	200	310	440	266.7
Natural gas (mcm)	7	7	3	8	14.3
Pig iron (tonnes '000)	250	980	1,450	1,930	672.0
Rolled steel (tonnes '000)	130	370	670	1,060	715.4
GVIO (bn. current yuan)	14.0	9.1	26.4	34.9	149.3
Light industry (bn. current yuan)	10.3	13.5	17.9	22.5	118.4
Heavy industry (bn. current yuan)	3.7	5.6	8.5	12.4	235.1

Notes: mcm million cubic metres. GVIO gross value of industrial output.

[a] For agricultural use, includes nitrogenous and phosphate fertilizers.

Source: State Statistical Bureau (1981–92 (1991), Beijing).

The liberal policies of the newly established government were apparently favourable to China's industrial sector between 1949 and 1952, as is evident from Table 3.1. During these years, the GVIO of China rose sharply from 14.0bn. yuan to 34.9bn. yuan. The biggest increase in value occurred in the heavy industrial sector of China, which rose in value over 235 per cent between 1949 and 1952, as compared to just over 118 per cent for the light industrial sector. Note, however, that heavy industry grew from an extremely small base, and relatively high prices of heavy industrial products led to an overestimation of growth rates.

In spite of the many political and social problems during the period of rehabilitation, production of pig iron and rolled steel increased greatly, by 672.0 per cent and 715.4 per cent respectively. This profound recovery of the heavy industrial sector was largely a result of incentives to workers and owners and, as previously outlined, general liberal policies (such as allowing private ownership to persist) towards the industrial sector. Other factors which contributed to the recovery included the organization of national unions to oversee factory services, the halting of inflation and balancing of the state budget by 1950, a simplification of taxation policies, and a sale of bonds.

Although there was a strong recovery in pig iron and rolled steel production between 1949 and 1952, many parts of China's industrial sector were still plagued by the Korean conflict and damage caused by the Civil War. However, rehabilitation of the overall economy was far enough along to prompt the Chinese government to plan for long-term economic development, with a focus on large-scale industrialization. After choosing to follow the Soviet model of development, the Chinese government implemented the First Five-Year Plan for economic development, which was officially adopted on 30 July 1955 but which effectively began in 1952.

THE FIRST FIVE-YEAR PLAN (1953–1957)

The first phase of China's development (1949–52) was a period of rehabilitation and recovery. In industry, full-scale nationalization or public ownership of the means of production was avoided so that experienced non-Communist technicians and workers could continue working. In late 1952, the Chinese government decided to undertake long-term economic planning, and adopted the Soviet model of development. At that time, Chinese leaders also realized that only the former Soviet Union was willing and able to provide financial and technical assistance to their country.

One principal feature of the Soviet model is its emphasis on the development of capital-intensive industries, particularly those producing raw materials, machinery, and other investment goods. During the First Five-Year Plan a socialist transformation of capitalist industry and commerce was successfully completed in China, although at a cost. The imbalance between large production units and small and middle-sized industrial units, between heavy and light industry, between developed and undeveloped areas, and between cities and rural areas became more pronounced after 1953 (Chesneaux 1979: 65). None the less, the fundamental goal of the First Five-Year Plan of establishing a preliminary foundation for the socialist industrialization of the national economy was attained (State Statistical Bureau 1960: 80). During the period 1953–7, GVIO increased by 56.4 per cent, with an average annual increase of 9.4 per cent (see Table 3.2). Other sectors of China's economy were also improved, including culture, agriculture, and handicrafts. Growth performance during the First Five-Year Plan period included restored production which was previously lost during the war years; only some increase in output represented actual new production.

To advance socialist transformation during the First Five-Year Plan, priority was given to the development of heavy industry. More than 86bn. yuan were invested in capital construction in China from 1952 to 1958, of which 22.4 per cent was for heavy industry (State Statistical Bureau 1960: 45–7). Between the same years more than 50,000 factory and mining construction projects were completely or partly built and put into operation (State Statistical Bureau 1960: 45–7). Remarkably, during the First Five-Year Plan, one large factory or mining construction project went into operation every three days on the average (State Statistical Bureau 1960: 47). One of the largest factory enterprises that was completed between 1953 and 1957 was the Anshan Iron and Steel works which, at that time, became one of the

TABLE 3.2. *Estimated production of selected industrial items during the First Five-Year Plan (1953–1957)*

	1953	1954	1955	1956	1957	% increase 1953–7
Cement (tonnes '000)	3,880	4,600	4,500	6,390	6,860	76.8
Chemical fertilizers[a] (tonnes '000)	50	67	79	111	151	202.0
Coal (m. tonnes)	70	84	98	110	131	87.1
Crude oil (tonnes '000)	620	790	970	1,160	1,460	135.5
Natural gas (mcm)	11	15	17	26	70	534.6
Pig iron (tonnes '000)	2,230	3,110	3,870	4,830	5,940	166.4
Rolled steel (tonnes '000)	1,470	1,720	2,160	3,140	4,150	182.3
GVIO (bn. current yuan)	45.0	51.5	53.4	64.2	70.4	56.4
Light industry (bn. current yuan)	28.2	31.7	31.6	37.0	38.7	37.2
Heavy industry (bn. current yuan)	16.8	19.8	21.8	27.2	31.7	88.7

Notes: mcm million cubic metres. GVIO gross value of industrial output.

[a] For agricultural use, including nitrogenous and phosphate fertilizers.

Source: State Statistical Bureau (1981–92 (1991), Beijing).

ten largest steelworks in the world, with an annual production capacity of more than 3m. tonnes.

The massive programme of socialist industrialization between 1953 and 1957 brought about large-scale increases in production capacities. Table 3.3 lists the increased production capacity of several major industrial products during the period, calculated according to the designed annual production capacities. According to China's State Statistical Bureau (1960), the rapid increase of production capacity evident in Table 3.3 was not possible in pre-1949 China. As evidence of this, it took 60 years prior to 1949 to develop a production capacity of only 1m. tonnes of steel. Steel-making capacity from 1949 to 1958 increased to more than 11 times that of pre-1949 China.

The increase in production capacity of steel-rolling, coal-mining, and petroleum production coincides with the dramatic rise in output of these commodities, as was shown in Table 3.2. Rolled steel, coal, and crude oil production increased 184.3 per cent, 87.1 per cent, and 135.5 per cent respectively between 1953 and 1957. Among the major metals, non-metals, and mineral fuels listed in Table 3.2, only production of cement increased at less than 80 per cent (approximately 76.8 per cent) during the First Five-Year Plan. Such dramatic rises in output of major industrial items attest to the Chinese government's commitment to the development of its heavy industrial sector during this period.

TABLE 3.3. *Newly increased production capacities, 1953–1957*

Major industrial product	Increased production capacity	
Electric power (installed capacity)	24,469	kilowatts
Coal-mining	63,760,000	tonnes
Petroleum	1,312,000	tonnes
Synthetic oil	522,000	tonnes
Iron-smelting	3,390,000	tonnes
Steel-making	2,820,000	tonnes
Steel-rolling	1,650,000	tonnes

Source: State Statistical Bureau (1960).

As for geological prospecting in China during the First Five-Year Plan, much success was achieved. The use of mechanical core drilling for prospecting escalated during the early 1950s, improving the success ratio of exploration activities. Between 1950 and 1958 the amount of drilling work that was conducted in the nation was estimated to be more than 130 times the total amount completed in nearly half a century before 1949 (State Statistical Bureau 1960: 49).

THE SECOND FIVE-YEAR PLAN (1958–1962)

The draft of China's Second Five-Year Plan was submitted to and approved by the Eighth National Party Congress in September 1956 and published, but its implementation was superseded by the launching of the Great Leap Forward in 1958. Five major tasks of the Plan were translated from Premier Zhou Enlai's proposals for the Second Five-Year Plan, delivered in September 1956 at the Eighth Chinese Communist Party National Congress (Liu Suinian and Wu Qungan 1986). According to the report, the fundamental tasks of the Plan were to: (1) carry out industrial construction centring around heavy industry, speed up the technical transformation of the national economy, and lay a solid foundation for China's socialist industrialization; (2) accomplish the task of socialist transformation, consolidate and expand ownership by the collective and ownership by the whole people; (3) further develop industrial, agricultural, and handicraft production and appropriately develop transportation and commerce; (4) make great efforts to train qualified personnel for national construction and upgrade scientific research work; and (5) strengthen national defence and improve the people's material and cultural well-being.

The stated proposals set forth the following major targets for the five years: total industrial and agricultural output value was to increase by about 100 per cent and

35 per cent respectively; steel output was to reach 10.5–12m. tonnes by 1962; the proportion of capital construction investment in the state's total revenue was to increase from 35 per cent in the First Five-Year Plan to about 40 per cent; the total amount of investment in capital construction was to increase by about 100 per cent; and the average income of the workers and farmers was to increase by 25 per cent to 30 per cent.

Although still largely based on the Soviet model, the Second Five-Year Plan actually contained important departures from or improvements upon the Soviet-type First Five-Year Plan. The Second Five-Year Plan attempted to introduce a more balanced strategy by placing more stress on smaller and less capital-intensive industrial plants, on agriculture and light industry, and on greater powers for local government in planning and economic decision-making.

While outlined by Premier Zhou in 1956, the Second Five-Year Plan was shortly abandoned for a more energetic movement—the Great Leap Forward (1958–60). The Great Leap did not so much discard the Second Five-Year Plan as try to realize the plan targets in three years instead of five. By the beginning of 1958 Chairman Mao Zedong and other Chinese leaders were reluctant to continue an economic strategy based on the Soviet model, as economic and political problems loomed on the horizon. Although the First Five-Year Plan was considered a success, imbalances grew between different sectors of the Chinese economy and geographic regions of the country. The Great Leap Forward was instituted in early 1958 as a means of greatly accelerating economic growth and advancing socialist construction according to China's specific economic conditions.[1]

Accompanying this sharply accelerated and over-ambitious timetable were important institutional reforms (for example, the People's Commune Movement, major decentralization of planning, moral incentives, backyard steel furnaces, local and rural industries, and so on) which made the Great Leap Forward distinctive from both the Soviet model and the original Second Five-Year Plan. High targets were set for industry, agriculture, transportation, and commerce as well as cultural, educational, and public health undertakings. All sectors were expected to achieve dramatic advances, and the iron and steel industry was to take the lead by doubling its output. This is extensively discussed in Riskin (1987) and Liu Suinian and Wu Qungan (1986).

The Great Leap Forward commenced when the CCP hung out the 'three red banners'—which represented the general line,[2] the people's communes, and the Great Leap Forward of production. The latter policy was adopted as Chairman Mao believed that industrial production in China could increase rapidly without delay. One means of achieving this objective was through total mobilization of China's labour-force. In this context, in April 1958 the People's Commune Movement started—an unparalleled institutional change in Chinese history. The communes were to be multi-purpose units responsible for managing industrial, agricultural, commercial, cultural, and military affairs (Donnithorne 1967). The communes were to combine political and economic functions and be both basic economic units and units of state power. Within a few months, an estimated 740,000 agricultural co-operatives were merged and reorganized into 26,000 large-scale people's communes which functioned as both productive units and local governments. The people's communes involved an estimated 120 million peasant households, or approximately 99 per cent of all peasant families in China (State Statistical Bureau 1960), and thus represented a form of social organization that was extremely significant to the economic development of China at that time.

Despite Chinese statements (see, for example, State Statistical Bureau 1960) that the Great Leap Forward and People's Commune Movement were unparalleled forms of organization for accelerating China's socialist economic transformation, the two strategies were unsuccessful and caused havoc in China's economy, particularly the industrial sector. The Great Leap Forward ignored the many constraints imposed by China's low level of economic development and the excessive waste of resources by the creation of a small-scale backyard industry, while the People's Commune Movement immediately led to work slow-downs and turmoil in rural areas. By 1960 it was apparent to the Chinese leadership that the two movements were failing, and economic readjustment was therefore begun in 1961.

China's leadership had recognized impending failures of the Great Leap Forward as early as mid-1959, as attested by the highly contentious party conference held at Lushan in Jiangxi province. For that meeting, Mao had drawn up a list of 19 problems arising from the leftism of the leap that remained to be dealt with. He also issued several directives calling each county and commune to compile lists of party faithfuls and cynics. But attempts to rectify excesses of the Great Leap Forward movement were overtaken by Mao's decision to attack Defence Minister Pen Te-huai, who had sharply criticized Mao and other leaders for launching the Great Leap (MacFarquahar 1983).

During the 1958 to 1960 period overall industrial output in China continued to rise in value terms. Table 3.4 shows that China's GVIO increased to nearly 164bn. yuan in 1960. The growth in output of major mineral and energy commodities during the same years represented continued uninterrupted growth started during China's First Five-Year Plan period.[3] However, with the deterioration of China's economy by 1960, minerals and energy output began to drop significantly as the harmful effects of the Great Leap Forward and People's Commune Movement gripped the industries.

As seen in Table 3.4, rolled steel production in China grew to nearly 11.1m. tonnes in 1960; yet to meet assigned quotas old-fashioned steel furnaces were used, resulting in a significant amount of poor-quality and even useless steel. In 1958, for example, hundreds of thousands of small blast furnaces were built for producing steel, but were operated by inexperienced workers utilizing outdated production processes. In actuality, the usable rolled steel produced between 1958 and 1960 was less than reported.

TABLE 3.4. *Estimated production of selected industrial items during the Second Five-Year Plan (1958–1962)*

	1958	1959	1960	1961	1962	% increase (decrease) 1958–62
Cement (tonnes '000)	9,300	12,270	16,650	6,210	6,000	(35.5)
Chemical fertilizers[a] (tonnes '000)	194	226	405	297	464	139.2
Coal (m. tonnes)	270	369	397	278	220	(18.5)
Crude oil (tonnes '000)	2,260	3,730	5,200	5,310	5,750	154.4
Natural gas (mcm)	110	290	1,040	1,470	1,210	1,000.0
Pig iron (tonnes '000)	13,690	21,191	27,160	12,810	8,050	(41.2)
Rolled steel (tonnes '000)	5,910	8,970	11,110	6,130	4,550	(23.0)
GVIO (bn. current yuan)	108.3	148.3	163.7	106.2	92.0	(15.1)
Light industry (bn. current yuan)	50.3	61.6	54.7	45.1	43.4	(13.7)
Heavy industry (bn. current yuan)	58.0	86.7	109.0	61.1	48.6	(28.9)

Notes: mcm million cubic metres. GVIO gross value of industrial output.

[a] For agricultural use, includes nitrogenous and phosphate fertilizers.

Source: State Statistical Bureau (1981–92 (1991), Beijing).

China's iron and steel industry was simply not capable of accomplishing what was required of it at this time. Examining iron-smelting, for example, shows that at the end of 1957 the annual production capacity of pig iron was 6.96m. tonnes. In the first eight months of 1958, only 5.3m. tonnes of pig iron were produced, meaning that 11.5m. tonnes more had to be turned out in the four months from September to December in order to fulfil the year's 'leap forward' plan (Liu Suinian and Wu Qungan 1986). As for mining and steel-rolling, the gaps were even wider between their production capacity and the targets set by the state. In the dramatic effort to boost iron and steel production, tremendous amounts of raw materials, labour, and money were wasted. Damage to heavy machinery was widespread as a result of over-capacity usage. A mass mobilization of nearly 50 million people to work in the iron ore mines and furnaces drew too much labour-force from China's countryside, contributing to a smaller than expected agricultural harvest in 1958.

After 1960, because of excessive wear and tear on machinery and transport equipment, steel output declined sharply (to just more than 4.5m. tonnes in 1962). Chairman Mao realized that in order to build large, modern steel mills, enormous capital investment was required. In order to avoid such investment, the Great Leap Forward attempted to substitute labour for capital using outdated technology, as China's growing population was considered a major resource to depend upon in creating large-scale investment projects necessary to address China's development problems. Mao's philosophy of extensive growth overnight resulted in the CCP establishing completely unattainable steel production targets. Millions of peasants joined the steel industry in China during early 1958, boosting the number of non-professional workers to 50 million, or 18.8 per cent of the total labour-force, in October of 1958 (US International Trade Commission 1985: 12). As a consequence, in order to meet the state's quotas unskilled steel workers produced low-quality steel using outdated technology. In some areas, where even large quantities of low-quality steel were not sufficient to meet quotas, exaggerated production figures were reported. Problems caused by unrealistic and in some cases unattainable quotas were not restricted to the steel industry, but affected other industries as well as the agricultural sector.

Another development which exacerbated problems in China's industries, and particularly steel, was the unexpected withdrawal of Soviet technical assistance in 1960. The problems generated were numerous and included poor management at steel mills and lack of skills in maintaining existing technology. An immediate impact on steel production was felt as a result of the Soviet withdrawal when the Anshan and Wuhan steel complexes were impaired, and closed down, respectively (Chesneaux 1979: 101).

China's First Five-Year Plan was successful, in part, because of Soviet assistance in nearly 156 turnkey investment projects for which the Soviets provided designs, equipment, engineers, and loans (Rawski 1979: 51). However, Sino-Soviet tensions grew stronger after 1953 primarily over Chairman Mao's periodic criticisms of the Soviet model of development. These tensions were sharply exacerbated after 1958 when Khrushchev, against Mao's advice, announced his policy of 'peaceful coexistence' with the capitalist West. Imbalances in the First Five-Year Plan led to a speech by Mao in 1956 where he enumerated the main imbalances in Chinese society (Mao Zedong 1977). This speech, although not publicized at the time, was an indirect critique of the (Soviet) economic philosophy of the First Five-Year Plan. When the Soviets withdrew from China in 1960, Chinese authorities were apparently already resolved to taking a more independent, self-reliant attitude that, in actuality, prevailed to the late 1970s.

THE 1961–1965 PERIOD OF RECOVERY AND READJUSTMENT

With the havoc caused by the Great Leap Forward, People's Commune Movement, and Soviet withdrawal, the Chinese government realized that its strategies for accelerated growth in heavy industry were impractical and unsuccessful. As a result, in January 1961 the CCP Central Committee formulated four guidelines for economic recovery—adjustment, consolidation, reinforcement or enrichment of weak links, and raising standards. These guidelines accurately depicted China's new economic strategy, and involved readjustment of the growth rate as well as intersectoral relations (under the guideline of adjustment the sectoral priorities of heavy industry, light industry, and agriculture were reversed), consolidation of attainable output levels and reorganization of the commune systems, a strengthening of weak links in the economy, and improvement in production technology and the quality of products. With the emergence of this strategy in 1961, the Chinese government recognized the constraining role of agriculture in China's development (Riskin 1987). By 1962 an 'agriculture first' policy was officially in effect in China as the expansion of agriculture was considered the starting-point of economic reconstruction. Agriculture continued, however, to be subordinated to industry, in both investment funds and pricing policy.

For purposes of this book the 1961–5 period is broken down into two sub-periods—a Period of Recovery (1961–2), and a Period of Readjustment (1963–5). The four guidelines proposed by the CCP were formulated in early 1961, and set the standards for restoration and readjustment of China's economic system. Because of the lagged effects of economic policies on actual industrial output levels, the industrial recovery from the economic difficulties of the 1958 through 1960 years was not really noticeable until 1963. By 1965, the last year of the five-year Period of Recovery and Readjustment, agricultural and industrial production reached levels which were targets established in the proposals for the Second Five-Year Plan. The Chinese economy had essentially fully recovered.

Although there was a new emphasis for developing the agriculture sector, the mining and metallurgical industries of China were nevertheless subject to expansion in the early 1960s. This was required, in part, to fulfil the needs for self-sufficiency in industrial raw materials and agricultural fertilizer resources. As the importance of agricultural development was stressed by the CCP in 1961, production of chemical fertilizer raw materials expanded throughout the Period of Recovery and Readjustment. The steel industry was even redirected towards making steel parts for agricultural equipment, and thus began producing increasing varieties of products between 1963 and 1965.

The Chinese government established many objectives for the nation's mining and industrial enterprises for the years of recovery and readjustment, including (1) improving product quality, (2) diversifying products, (3) reducing cost and increasing productivity, (4) conserving materials and recovering more from waste, and (5) developing new mineral and energy reserves (US Bureau of Mines 1951–92 (1963–5): iv). To improve efficiency in mining and processing, many mines and plants were consolidated and revamped, and successful small mines and plants were slated for expansion. Owing to the withdrawal of Soviet technical assistance in 1960, efforts were made by the Chinese government in later years to train Chinese technicians in facilities staffed by Chinese, and in some cases designed by them. Although there were increased efforts to build mining facilities of Chinese design, foreign technology was still being purchased to 1965.

Table 3.5 lists production figures for several major metals, non-metals, and mineral fuels in China during the Period of Readjustment (1963–5). It is evident that the policies and programmes adopted during the 1961–2 Period of Recovery brought about noticeable improvements in China's economy and increased industrial and minerals/energy development. Significantly increased production of pig

TABLE 3.5. *Estimated production of selected industrial items during the Period of Readjustment (1963–1965)*

	1963	1964	1965	% increase 1962–5
Cement (tonnes '000)	8,060	12,090	16,340	102.7
Chemical fertilizers[a] (tonnes '000)	648	1,008	1,726	166.4
Coal (m. tonnes)	217	215	232	6.9
Crude oil (tonnes '000)	6,480	8,480	11,310	74.5
Natural gas (mcm)	1,020	1,060	1,100	7.8
Pig iron (tonnes '000)	7,410	9,020	10,770	45.3
Rolled steel (tonnes '000)	5,330	6,880	8,810	65.3
GVIO (bn. current yuan)	99.3	116.4	140.2	41.2
Light industry (bn. current yuan)	44.5	51.6	72.3	62.5
Heavy industry (bn. current yuan)	54.8	64.8	67.9	23.9

Notes: mcm million cubic metres. GVIO gross value of industrial output.

[a] For agricultural use, includes nitrogenous and phosphate fertilizers.

Source: State Statistical Bureau (1981–92 (1991), Beijing).

iron and rolled steel between 1963 and 1965 was possible through application of new technology in existing facilities, streamlining operations, continued construction of partially completed steel centres, employing more coal-blending to conserve coking coal for blast furnace operations, and introducing oxygen converters to complement the open hearths (US International Trade Commission 1985). The newly established sectoral priorities of agriculture, light industry, and heavy industry helped stimulate a dramatic increase in output of chemical fertilizers of more than 166 per cent between 1963 and 1965. More than 1.7m. tonnes of chemical fertilizers were produced in China by 1965, with 60 per cent being nitrogenous in form and the remainder being phosphate.

Unlike some mineral or energy resources in China, crude oil production was not slowed by the turmoil caused by the Great Leap Forward, People's Commune Movement, and Soviet withdrawal, in large part owing to its importance to the nation's overall economy. As seen in Table 3.5, output of crude oil increased nearly 75 per cent during 1963–5, which was a continuation of the rising output begun during the Second Five-Year Plan. The rate of increase in production did slow, however, during the Period of Recovery (1960–2), but none the less was positive. By 1965 China was basically self-sufficient in oil, with development of the Daqing field being largely responsible.

Coal output during the Period of Readjustment increased about 7 per cent. Such an increase was not as significant as other major mineral or energy products during the period, suggesting that coal production did not recover fully from the problems caused by the Great Leap Forward. Indeed, a dramatic decrease in coal output occurred immediately following the Great Leap Forward, from 397m. tonnes (1960) to 215m. tonnes (1964). This drop forced the closure of many inefficient small mines in China, and disrupted operations at large mines. It was not until 1962 and 1963 that efficient small mines were consolidated and expanded, and operations of large existing mines became normalized (US Bureau of Mines 1951–92 (1964): iv. 1166). Production of coal soon rebounded to 232m. tonnes in 1965.

THE THIRD FIVE-YEAR PLAN (1966–1970)

Not much was ever reported on the contents of China's Third Five-Year Plan until recently (Liu Suinian and Wu Qungan 1986). Economic documents and statistics of China published after 1959 and before 1978 were scarce, though detailed annual output series for this period are also available in the various statistical yearbooks published since 1979. During the 1966 to 1970 period, the growth rates of industrial and agricultural output in China were higher than in the previous eight years, though markedly lower than those of the First Five-Year Plan period (Xue Muqiao 1982).

The Cultural Revolution of China—officially known as the Great Proletarian Cultural Revolution—was an unprecedented phenomenon in the history of Chinese communism. The revolution was primarily a struggle for power in which Mao's underlying goal was to strengthen his leadership over the party, army, government, and 'reactionary bourgeois authorities' in control of the schools and cultural institutions. It aimed at building China into a genuinely socialistic nation without bureaucracy, élitism, or inequalities (Su Kaiming 1986). Many measures and policies similar to those adopted during the Great Leap Forward were reinstated between 1966 and 1970, leading to adverse economic and industrial conditions.

Two major outcomes of the Cultural Revolution included the activation of Maoist militant students known as Red Guards (*hongweibing*), and the use of the People's Liberation Army (PLA). The Red Guards demonstrated against power holders in China's education system, while the PLA was used primarily to stop armed clashes between different Red Guard factions. The Red Guards attacked scholars and professors, and fighting often broke out between the militant students and peasants or industrial workers in various parts of China. These activities eventually contributed to a disruption of the nation's industrial system.

Table 3.6 reveals this disruption, particularly during the 1966 to 1968 period when the Red Guards were especially active. In 1967 industrial output value fell by 14.9 per cent, to 138.2bn. yuan. It fell another 7 per cent in 1968. During the same two-year period rolled steel output dropped nearly 36 per cent, from 10.4m. tonnes in 1966 to 6.7m. tonnes in 1967. Crude oil production fared relatively well

TABLE 3.6. *Estimated production of selected industrial items during the Third Five-Year Plan (1966–1970)*

	1966	1967	1968	1969	1970	% increase 1966–70
Cement (tonnes '000)	20,150	14,620	12,620	18,290	25,750	27.8
Chemical fertilizers[a] (tonnes '000)	2,409	1,641	1,109	1,749	2,435	1.1
Coal (m. tonnes)	252	206	220	266	354	40.5
Crude oil (tonnes '000)	14,550	13,880	15,990	21,740	30,650	110.7
Natural gas (mcm)	1,340	1,460	1,400	1,960	2,870	114.2
Pig iron (tonnes '000)	13,340	9,630	8,570	12,800	17,060	27.9
Rolled steel (tonnes '000)	10,350	7,180	6,660	9,260	11,880	14.8
GVIO (bn. current yuan)	162.4	138.2	128.5	166.5	211.7	30.4
Light industry (bn. current yuan)	79.6	73.3	69.0	83.7	97.6	22.6
Heavy industry (bn. current yuan)	82.8	64.9	59.5	82.8	114.1	37.8

Notes: mcm million cubic metres. GVIO gross value of industrial output.

[a] For agricultural use, includes nitrogenous and phosphate fertilizers.

Source: State Statistical Bureau (1981–92 (1991), Beijing).

during the first three years of the Cultural Revolution, resulting largely from increasing output at Daqing and the newly developed Shengli oil field. And importantly, the Chinese leadership had instituted a policy which barred politically led disruptions from entering and disrupting the nation's oil fields, where business essentially proceeded as usual during the mid-1960s. Protection of the nation's oil industry at this time was considered crucial, as China's government realized the economic significance of crude oil development to the country's ambitions for advancing socialist construction.

China's coal industry was particularly hard hit by the political and social disturbances in 1967, when output dropped more than 18 per cent from the preceding year (Table 3.6). In September 1967 Premier Zhou Enlai had warned demonstrators in Beijing that a serious shortfall of coal would develop in the following winter if workers did not return to their coal-mines and boost production.[4] During that year the Chinese government instituted rationing policies for coal (for private consumption) as a result of decreased coal output. The drop in coal production between 1966 and 1967 created an energy deficiency which affected the output of several other industries, such as iron and steel.

Industrial production in China rebounded in 1969 and 1970, although figures were probably inflated and did not reveal some of the lasting damaging impacts of the recent unrest on the nation's industrial system. In September 1967, through a directive, Chairman Mao had authorized the People's Liberation Army to stop all Red Guard activities, as the students had outlived their usefulness and were uncontrollable. Upon receiving this authority the People's Liberation Army was given unprecedented access to management of the economy, including minerals and energy production (US Bureau of Mines, 1951–92 (1968): iv. 189). This action was evidently favourable to China's mining industry as output grew strongly once again in 1969 and 1970, though much of the growth included restored production. Overall, economic conditions improved when the People's Liberation Army was in control, particularly in late 1969 and 1970 when the Army brought factional disputes under control. Efforts were made to stimulate industrial production when in late 1969 a nationwide emulation campaign was begun at the Peking Steelworks, which was to serve as a model industrial enterprise. This campaign contributed to a growing feeling of optimism in China about the state of the nation which, in turn, helped China's economy grow. Gross industrial output value rose dramatically between 1968 and 1970, by nearly 62 per cent. At the same time rolled steel output increased 15.8 per cent, coal output by 60.9 per cent, and crude oil output by 91.7 per cent.

THE FOURTH FIVE-YEAR PLAN (1971–1975)

The Cultural Revolution lasted over ten years, from May 1966 to October 1976 (Su Kaiming 1986), although the most severe effects were felt over a considerably shorter period. Su Kaiming (1986) has divided the Cultural Revolution into three stages, where the first and very turbulent stage covered roughly three years, starting with the '16 May Circular' in 1966 in which Mao pointed out the necessity for the Cultural Revolution, and ending with the convening of the Ninth National Congress of the Party in April 1969 which named Lin Biao as Mao Zedong's successor. The second period—progress of the revolution and Lin Biao's plot to assassinate Mao Zedong and its failure—covered the period from May 1969 to August 1973, the interval between the ninth and tenth congresses of the Party. During the third stage, which lasted from the conclusion of the Tenth Party Congress in August 1973 to the downfall of the Gang of Four in October 1976, Jiang Qing and her associates attempted repeatedly to grasp the supreme political power for themselves, but failed.

As with the Third Five-Year Plan, much of the contents of the proposed Fourth Five-Year Plan were not published until recently (Liu Suinian and Wu Qungan 1986). The Cultural Revolution continued through the Fourth Five-Year Plan period, although the disruptive first stage had ended in 1969. Economic and industrial growth in China between 1971 and 1975 was generally positive but relatively erratic. A small-scale 'great leap forward' occurred in 1971–2, leading to declines in output growth rates in 1974. There were sharp annual fluctuations in economic prosperity during the Fourth Five-Year Plan period.

By 1971 China's gross industrial output value had risen to 237.5bn. yuan (Table 3.7), representing an increase of 46.2 per cent over the 1966 figure and 84.8 per cent over 1968. After 1971 output continued to increase steadily at an average annual rate of 7 per cent to 1975. Much of the growth in China's industrial output

TABLE 3.7. *Estimated production of selected industrial items during the Fourth Five-Year Plan (1971–1975)*

	1971	1972	1973	1974	1975	% increase 1971–5
Cement (tonnes '000)	31,580	35,470	37,310	37,090	46,260	46.5
Chemical fertilizers[a] (tonnes '000)	2,994	3,701	4,592	4,222	5,247	75.3
Coal (m. tonnes)	392	410	417	413	482	23.0
Crude oil (tonnes '000)	39,410	45,670	53,610	64,850	77,060	95.5
Natural gas (mcm)	3,740	4,840	5,980	7,530	8,850	136.6
Pig iron (tonnes '000)	21,000	23,550	24,900	28,620	24,490	16.6
Rolled steel (tonnes '000)	13,890	15,610	16,840	14,660	16,220	16.8
GVIO (bn. current yuan)	241.4	256.5	279.4	279.2	320.7	32.9
Light industry (bn. current yuan)	103.7	110.0	121.2	124.1	141.3	36.3
Heavy industry (bn. current yuan)	137.7	146.5	158.2	155.1	179.4	30.3

Notes: mcm million cubic metres. GVIO gross value of industrial output.
[a] For agricultural use, includes nitrogenous and phosphate fertilizers.
Source: State Statistical Bureau (1981–92 (1991), Beijing).

during this period was fuelled by a substantial increase in minerals and energy production, particularly of chemical fertilizers and crude oil.

The 'agriculture first' policy begun during the Period of Recovery (1961–2) was partly continued through the Fourth Five-Year Plan years. Agriculture was again, however, strongly subordinated to industry in real economic terms. Economic surplus from the agricultural sector was used to fund China's continuing industrialization drive. While nominally the policy was 'agriculture first', during the Fourth Five-Year Plan period, investment and pricing strategies continued to be biased towards industry.

Between 1971 and 1975, the end of the Fourth Five-Year Plan, it is estimated that chemical fertilizer production rose nearly 75 per cent. From 1970 to 1973 the number of tractors and diesel engines supplied to agriculture was reportedly greater than that of the preceding 20 years (Cheng 1982). In terms of investment in capital construction between 1971 and 1975, however, agriculture still received less than 10 per cent of the total, compared to 49.6 per cent to heavy industry.

In addition to increased chemical fertilizers output and construction of agricultural machinery in the early 1970s, efforts were made in China to support agriculture through an expansion of basic industry. To facilitate industrial expansion, emphasis was placed on making the tools of production, particularly mining, beneficiation, metallurgical, chemical, and refining equipment and plants (US Bureau of Mines 1951–92 (1972): iii. 221). Chinese rolled steel production rose from a low of 6.7m. tonnes in 1967 to nearly 16.2m. tonnes in 1975. Such an increase in production was, in part, possible because of a steadily rising output of iron ore and other ingredients used in the making of steel. This follows the strong effort on the part of the Chinese government at that time to expand the nation's iron ore base in order to meet the needs of an expanding steel industry. The pressure to substantially expand China's steel industry increased towards 1975 as a result of a trend towards greater industrialization to sustain the economy (K. P. Wang 1975: 49).

Two new economic philosophies were brought forth during the Fourth Five-Year Plan period—the four modernizations and the open-door policy. In 1971, Chairman Mao turned over responsibility for accelerating development of China's economy to Premier Zhou, who realized that to modernize the nation's industrial system foreign technology would have to be imported. With Mao's support, Zhou Enlai worked towards reviving China's economy, restoring order, and improving relations with foreign countries. As a result, 1971 was characterized by a thawing of relations between China and the Western world, and particularly the United States. In February 1972, United States President Nixon paid a historic visit to China and signed the Shanghai Communiqué, which provided a framework for China to import US industrial technology (US International Trade Commission 1985: 22). By 1972 Premier Zhou had worked with Deng Xiaoping to rejuvenate the Chinese economy and stimulate trade with foreign nations. On the notion that an outward-looking strategy was needed, Zhou prepared his four modernizations proposal to develop China's economy through modernization of agriculture, industry, science and technology, and defence.[5]

Despite Zhou's support of a four-modernizations and open-door programme, in the early 1970s China did not actively seek or consider participation in industrial or minerals-related joint ventures with foreign nations. The isolationist tendencies of previous years tended to prevail throughout the Chinese mining industry. At that time the 'Gang of Four'—a radical faction committed to the policies of the Cultural Revolution—were vehemently opposed to foreign participation in the exploitation of China's natural resources. Since there was, however, some need to acquire advanced mining technology from abroad, a few visits by Japanese and American officials were made to China and some foreign equipment was purchased under special conditions from these countries. The Japanese tours were perhaps the more successful, involving technological exchanges of steel equipment, fertilizers, metallurgy, and basic chemicals and petrochemicals. By 1974 and 1975 the Chinese government was also attempting to modernize small-scale inefficient mines through the purchase of foreign plants, equipment, machinery, and technology.

THE FIFTH FIVE-YEAR PLAN (1976–1980)

The principal goal of China's post-Cultural Revolution leadership was to recoup the economic losses suffered during the previous years of turmoil and accelerate the growth of China's economy. In 1977–8, Mao's temporary successor Hua Guofeng introduced a ten-year economic plan to overlay the two constituent Fifth and Sixth Five-Year Plans. The plan sought to build China into an independent industrial and economic system during the years 1976 to 1985, enabling the nation to achieve an industrial ranking on a par with other developed nations by the turn of the century. The plan was the centre of Hua's overall strategy to facilitate incredible modernization in China by 1985.

It is believed that a draft of the Fifth Five-Year Plan was prepared by Zhou and Vice-Premier Deng Xiaoping, although the plan itself was never published and rarely cited in Chinese sources. After Zhou died on 8 January 1976, Deng was vulnerable to criticism from radicals, including the newly selected Premier Hua Guofeng. In early 1978 Hua introduced a ten-year modernization programme, hoping that he would remain in power throughout the period. The Ten-Year Plan (1976–85) was prepared in accordance with Zhou Enlai's call for four modernizations and Mao's strategy of high-speed development. Specifically, the Plan called for agricultural production in China to reach or exceed levels of developed countries by the end of the century, and output of industrial goods to approach, equal, or surpass that of the developed world (*Peking Review*, 10 Mar. 1978: 18). In line with Zhou Enlai's ambitions, by the year 2000 China was to take its place in the 'front ranks' of the world.

In terms of industrial development, the Ten-Year Plan envisaged an advanced heavy industry with the metallurgical and mineral fuel industries being further developed through the adoption of new techniques, and the output of iron and steel, coal, and crude oil on par with the world's leading producing nations by 1985 (*Peking Review*, 10 Mar. 1978: 19). At the same time, China's petrochemical industry was to be greatly expanded.

To achieve these ambitious objectives, the Chinese government had planned to invest between the years 1978 to 1985 an amount equal to the total capital construction investment for the preceding 28 years. If compared to the First Five-Year Plan, the share of capital investment into industry dropped relative to the share in agriculture. This follows the endorsement of 'agriculture as the foundation of the national economy' included in the Ten-Year Plan. To support the economic expansion of China to 1985, the Plan called for increased foreign trade and investment in China where needed.

Hua's proposals outlined in his Ten-Year Plan directly impacted the mining industry of China. Specifically, steel continued to be regarded as the key link in China's industrial development, and efforts were planned to renovate and expand the industry. The state planned to build or complete 120 large-scale projects, which included ten iron and steel complexes, nine non-ferrous metal complexes, eight coal-mines, and ten oil and gas fields (*Peking Review*, 10 Mar. 1978: 22–3). The completion of these and other projects was to provide China with 14 fairly strong and rationally located industrial bases.

A distinctive feature of Hua's ten-year strategy was a very ambitious programme of crude oil exports to finance the massive investments and foreign technology imports of the building of new plants and equipment required for the plan. This unattainable oil export plan was based on over-optimistic assessments of recoverable reserves both onshore and offshore.

Within one year after Hua's announcement of the Ten-Year Plan in February 1978, nearly 100,000 construction projects of various sizes were begun (Liu Suinian and Wu Qungan 1986). Capital investment for agriculture, industry, transportation, and public utilities totalled 61,970m. yuan, or approximately $40bn., which represented a 50 per cent increase over the preceding year (Cheng 1982). However, owing to a lack of continuing capital funds as well as advanced technologies and materials, many of these projects had to be abandoned. The high economic and industrial growth rates called for in the Ten-Year Plan had proved to be impractical and unrealistic. Moreover, when problems were encountered Hua's call for greater efforts and more investment made the problems worse. At this time Premier Hua's faction lost control and Deng returned to power and called for an abandonment of the impractical policies of the Ten-Year Plan. As a result, in December 1978, a Three-Year Adjustment Plan (1979–81) was substituted for the existing Ten-Year Plan, marking the emergence of major economic, political, cultural, and social reforms.

Under the 1979–81 Three-Year Adjustment Plan, capital investment was slashed in several sectors, including the metallurgical, machine-building, and chemical industries. According to Xinhua News Service (15 Apr. 1980), in 1979 construction was halted of 384 large and medium-sized projects, the majority of which were in the metallurgical, machine-building, and chemical industries. China's capital outlays for the metallurgical industry were cut by 45 per cent in 1979. A decrease of investment in China's steel industry represented a fundamental shift in the nation's development priorities. The slack in capital investment was in turn picked up by the textile and light industries, as well as China's agricultural sector.

Because of the over-ambitious investment and construction targets called for in China's Ten-Year Plan, some human and capital resources were wasted and inefficiently utilized. However, though the rate of growth of output of a few industrial commodities slowed by the end of 1978, overall production continued to rise to 1980 (Table 3.8). China's GVIO reached approximately 490bn. yuan in 1980, a 20.4 per cent increase over 1978, and 55.1 per cent increase over 1976. The growth of light industry in China was particularly strong during the 1976 to 1980 period, increasing at an average annual rate of 13.4 per cent.

China's mining industry continued to expand during the late 1970s, following the endorsement of the Ten-Year Plan to advance heavy industry through major new construction and the employment of new techniques. The performance of the steel industry in 1978, 1979, and 1980 was outstanding, with rolled steel output rising roughly 23 per cent in three years. During this time, much of the idle steel

TABLE 3.8. *Estimated production of selected industrial items during the Fifth Five-Year Plan (1976–1980)*

	1976	1977	1978	1979	1980	% increase 1976–80
Cement (tonnes '000)	46,700	55,650	65,240	73,900	79,860	71.0
Chemical fertilizers[a] (tonnes '000)	5,244	7,238	8,693	10,654	12,321	135.0
Coal (m. tonnes)	483	550	618	635	620	28.4
Crude oil (tonnes '000)	87,160	93,640	104,050	106,150	105,950	21.6
Natural gas (mcm)	10,100	12,120	13,730	14,510	14,270	41.3
Pig iron (tonnes '000)	22,330	25,050	34,790	36,730	38,020	70.3
Rolled steel (tonnes '000)	14,660	16,330	22,080	24,970	27,160	85.3
GVIO (bn. current yuan)	327.8	372.5	423.7	468.1	515.4	57.2
Light industry (bn. current yuan)	144.8	163.8	182.6	204.5	243.0	67.8
Heavy industry (bn. current yuan)	183.0	208.7	241.1	263.6	272.4	48.9

Notes: mcm million cubic metres. GVIO gross value of industrial output.

[a] For agricultural use, includes nitrogenous and phosphate fertilizers.

Source: State Statistical Bureau (1981–92 (1991), Beijing).

capacity in China became more fully utilized. Gains in coal and crude oil production between 1978 and 1980 were nowhere near as strong as in steel, as the energy industry was concentrating much of its efforts on improving efficiency and catching up on development. By 1980 China's coal output reached third in the world, and crude oil sixth. Production of chemical fertilizers increased 135 per cent during the entire five-year period, continuing the rapid rise begun several years earlier. By 1980 12.3m. tonnes of chemical fertilizers were produced in China, 81.1 per cent being nitrogenous in form. Along with water projects and insecticides, output of chemical fertilizers in the 1970s was a fundamental element for advancing agriculture in China. Producing as much fertilizer as possible and in as many varieties as allowable was the fundamental policy that achieved great success.

THE SIXTH FIVE-YEAR PLAN (1981–1985)

In a change of attitude from earlier periods, the Chinese government encouraged foreign capital, technology, and ventures through an open-door policy to assist the nation's modernization drive during the Sixth Five-Year Plan. After the Ten-Year Plan was abandoned (late 1979), the State Planning Commission began to draft the Sixth Five-Year Plan for social and economic development of China (see Foreign Language Press 1984). This plan was finally approved by the Fifth Session of the Fifth National Congress on 10 December 1982. Prepared by Deng Xiaoping, Hu

Yaobang, and Zhao Ziyang, the Sixth Five-Year Plan paved the way for heavy industry to support growth and development in light industry and agriculture. Specifically, the Plan called for moderate growth rates in industrial production (average annual increase of 4 per cent), lower consumption of materials and energy in production, active development of relations with foreign countries (including imports of advanced technology), and technological transformation of existing enterprises based on assessments of priorities (*China Reconstructs* 1984).

Key elements in the shift of development strategy promoted in the Sixth Five-Year Plan included greater stress on (1) consumer goods to enhance incentives, (2) non-productive investments (for example, housing) to improve living standards, (3) participation in the international division of labour, (4) encouraging foreign direct investments and the creation of special economic zones, and (5) improving economic efficiency as a source of intensive growth (compared to the extensive growth of the past which was based primarily on huge growth of material inputs). The shift in sectoral priorities during the plan period affected relative growth rates and led to greater savings through higher efficiency levels. In agriculture, acceleration of reform occurred during the period with decollectivization or the introduction of the household production-responsibility system on a nationwide basis. This was accompanied by the parallel expansion of urban and rural open markets, a downsizing of state monopoly purchases, and a relaxation in the rationing of supplies (Lin 1989).

Systematic changes in China's non-agricultural sectors between 1981 and 1985 were intended to implement the policy of readjusting, restructuring, consolidating, and improving the national economy. A majority of the measures adopted during the period involved efforts to lower consumption of materials in production, particularly of energy consumption. In 1981, a production-responsibility system was introduced in industry, modelled after the agricultural counterpart. Its objective was to enhance microeconomic efficiency in the use of resources and to reduce financial losses through independent profit-and-loss accounting (Lin 1989). In 1983, a CCP Circular introduced measures for consolidating and improving enterprise management which paved the way for comprehensive industrial reforms on taxation, company reorganization, and specialization.[6]

While the iron and steel sector was de-emphasized during the Sixth Five-Year Plan period, the energy industry continued to receive high priority. Of the 120 projects designated for construction between 1981 and 1985, 22 supported an increase in energy output. These included costly renovations at the Daqing and Shanghai petrochemical complexes, and development or expansion of eight major coal-mining areas. In 1985 the State Statistical Bureau reported that a total of 70.5bn. yuan was spent on energy projects during the Sixth Five-Year Plan period, 25.6 per cent more than during the previous five years (*China Daily*, 18 Sept. 1985). Production capacity of coal-mines increased by 84.6m. tonnes, significantly more than what was targeted. Oil wells drilled and put into production between 1981 and 1985 reached an aggregate capacity of 44.9m. tonnes, 9.9m. tonnes over the target, while power stations added a total generating capacity of 19.6m. kw, 6.7m. kw over the goal.

Growth in China's energy output during the Sixth Five-Year Plan provides some evidence that increased investment in the sector may have had an impact, particularly in the later years of the period. Table 3.9 shows that coal and crude oil production increased 40.2 and 23.4 per cent respectively during the five years.

TABLE 3.9. *Estimated production of selected industrial items during the Sixth Five-Year Plan (1981–1985)*

	1981	1982	1983	1984	1985	% increase 1981–5
Cement (tonnes '000)	82,900	95,200	108,250	123,020	145,950	76.1
Chemical fertilizers[a] (tonnes '000)	12,390	12,781	13,789	14,602	13,222	6.7
Coal (m. tonnes)	622	666	715	789	872	40.2
Crude oil (tonnes '000)	101,220	102,120	106,070	114,610	124,900	23.4
Natural gas (mcm)	12,740	11,930	12,210	12,310	12,930	1.5
Pig iron (tonnes '000)	34,170	35,510	37,380	40,010	43,840	28.3
Rolled steel (tonnes '000)	26,700	29,020	30,720	33,720	36,930	38.3
GVIO (bn. current yuan)	540.0	581.1	646.0	761.7	971.6	79.9
Light industry (bn. current yuan)	278.1	291.9	313.5	360.8	457.5	64.5
Heavy industry (bn. current yuan)	261.9	289.2	332.6	400.9	514.1	96.3

Notes: mcm million cubic metres. GVIO gross value of industrial output.

[a] For agricultural use, includes nitrogenous and phosphate fertilizers.

Source: State Statistical Bureau (1981–92 (1991), Beijing).

Production of these two fuels was virtually at a standstill between 1978 and 1980. With the greater than 23 per cent increase in China's crude oil production between 1981 and 1985, output reached 125m. tonnes by the end of the period, with 44 per cent (55m. tonnes) coming from the Daqing oil field. Other fields or areas contributed percentages as follows: Shengli (21.6), Huabei (8), Liaohe (7.2), Zhongyuan (4.8), Xinjiang (4), and Dagang (3.2). In 1985, the output of crude oil, coal, and electricity rose between 7.7 and 8.9 per cent, compared with 1984.

In April 1985 China's leadership opened ten southern onshore areas to foreign oil exploration, keeping in line with the Sixth Five-Year Plan's strategy of increasing reliance on foreign capital and technology. The areas were located in Anhui, Fujian, Guangdong, Guangxi, Guizhou, Hunan, Jiangsu, Jiangxi, Yunnan, and Zhejiang. Prospecting and tapping offshore oil resources in such areas as the Bohai Sea and the Beibu Gulf in the South China Sea was also actively pursued in co-operation with foreign countries.

The decreased emphasis on the steel industry during the Three-Year Adjustment Plan period (1979–81) contributed to the slowing of output growth between

1982 and 1983. Table 3.9 shows that rolled steel production remained almost steady at between 29 and 30.7m. tonnes in 1982 and 1983. By 1984, however, rolled steel output was about 10 per cent greater than the previous year, reaching 33.7m. tonnes. By the end of the Sixth Five-Year Plan period China's iron and steel industry ranked fourth in the world after that of the Soviet Union, Japan, and the United States in order of tonnage; however, expansion of capacity was limited because of outdated, inefficient facilities and lack of high-grade iron ore resources.

According to the Sixth Five-Year Plan, production of phosphate and potash fertilizers for agricultural use was to be stepped up, and the proportion of phosphate and potash fertilizers in the total output of chemical fertilizers was to be increased gradually. As shown in Table 3.9, by 1985 13.2m. tonnes of chemical fertilizers were produced in China, up 6.7 per cent from 1981 and just short of the target output of 13.4m. tonnes. However, the increases in phosphate and potash fertilizers during the period did not reach planned levels, falling short by significant quantities. In contrast, output of nitrogenous fertilizer exceeded the planned target level, reaching 11.4m. tonnes in 1985. As a result the ratio between nitrogenous, phosphate, and potash fertilizers was not adjusted as intended during the period.

THE SEVENTH FIVE-YEAR PLAN (1986–1990)

The policy of opening to the international community promoted in the Sixth Five-Year Plan continued into the Seventh Five-Year Plan period, combined with the introduction of structural reforms for agriculture and industry, ambitious efforts to invigorate the domestic economy by the introduction of limited market mechanisms, and decentralization. The goal of modernization by the end of the century, as outlined by General Secretary Hu Yaobang in September 1982, 'to quadruple the gross annual value of industrial and agricultural products in the year 2000', remained intact.[7] To restore order to a troubled economy, a Three-Year Austerity Programme (1988–91) was introduced in late 1988 which outlined tough measures to revive state controls, curtail spending, and reduce imports of consumer goods. The plan, which was not officially released until January 1990, urged recentralization of control over trade and producing enterprises in China as a means of enhancing the state's role in guiding the economy. Agricultural and industrial reforms introduced early in the period were effectively put on hold when austerity policies were first adopted in autumn 1988 and particularly after the ouster of the pro-reform party leader Zhao Ziyang when pro-democracy demonstrations were ended in June 1989.

The period of the Seventh Five-Year Plan was a crucial one for the comprehensive reform of China's economic structure. A decade had passed since China began major reforms of its economy and political structure. During the period China joined the global economic system, and although the country's domestic economic reforms experienced curtailment beginning in 1988, they have been on a massive scale and instrumental in transforming the Chinese economy. Some level of reform will continue into China's Eighth Five-Year Plan period (1991–5) on an experimental basis and may reach into political, social, and cultural spheres of Chinese life.

During the Seventh Five-Year Plan period, China's national economy developed at an increased rate, and the Plan's targets were essentially fulfilled (Table 3.10). However, because of difficulties in introducing additional market-oriented reforms compounded by the impact of overheated economic growth towards the end of the Sixth Five-Year Plan period, the economy was in a state of flux.

China's national income rose at an average annual rate of 7.2 per cent during the period, exceeding the targeted growth rate of 6.7 per cent. Gross industrial

TABLE 3.10. *Estimated production of selected industrial items during the Seventh Five-Year Plan (1986–1990)*

	1986	1987	1988	1989	1990	% increase 1986–90
Cement (tonnes '000)	166,060	186,250	210,140	210,290	209,710	26.3
Chemical fertilizers[a] (tonnes '000)	13,597	16,722	17,402	18,025	18,797	38.2
Coal (m. tonnes)	894	928	980	1,054	1,080	20.8
Crude oil (tonnes '000)	130,690	134,140	137,050	137,640	138,310	5.8
Natural gas (mcm)	13,760	13,890	14,260	15,049	15,298	11.2
Pig iron (tonnes '000)	50,640	55,030	57,040	58,200	62,380	23.2
Rolled steel (tonnes '000)	40,580	43,860	46,890	48,590	51,530	27.0
GVIO (bn. current yuan)	1,119.4	1,381.3	1,822.5	2,201.7	2,392.4	113.7
Light industry (bn. current yuan)	533.0	665.6	897.9	1,076.1	1,181.3	121.6
Heavy industry (bn. current yuan)	586.4	715.7	924.5	1,125.6	1,211.1	106.5

Notes: mcm million cubic metres. GVIO gross value of industrial output.

[a] For agricultural use, includes nitrogenous and phosphate fertilizers.

Source: State Statistical Bureau (1981–92 (1991), Beijing).

output value reached 2,392.4bn. yuan by 1990, a 113.7 per cent increase over 1986, the first year of the period. Capital construction projects increased production capacity in many areas, including: 130m. tonnes of raw coal, 45.43m. kw in electricity generation, 63.56m. tonnes of petroleum, 5.12m. tonnes of steel, and 1.71m. tonnes of chemical fertilizers (*Beijing Review*, 7–13 Jan. 1991). China also recorded increases in imports and exports during the period, resulting in a readjustment in commodity mix. Total trade volume reached US$108bn., with the annual average growth rate of exports ending at 15.8 per cent, and imports somewhat lower at 3.8 per cent.

China's Seventh Five-Year Plan initiated several policies for geological prospecting activities for the period 1986–90, including speeding up mineral surveys and evaluation of reserves; boosting prospecting efforts for oil, natural gas, and coal and verifying new reserves of these if possible; and improving the general survey of gold, silver, copper, aluminium, lead, zinc, tin, sulphur, borax, and other mineral ores and locating additional exploitable reserves (Xinhua News Agency News Bulletin 1986: 24). Although nine metal and non-metal minerals are specifically mentioned in the Plan, four of them were subject to extensive exploration efforts—copper, lead, zinc, and bauxite (aluminium). Currently, China depends on imports for half of its copper consumption and this situation is expected to persist through the 1990s. Shortages of copper and copper alloys, zinc and zinc alloys, and aluminium and aluminium alloys led to increased imports of these metals into China during the past decade. Other minerals subject to increased exploration between 1986 and 1990 included tungsten and molybdenum, two important foreign exchange earners for China. China's State Statistical Bureau reported that 267 economically viable mineral deposits were discovered in China in 1990, while 102 mineral-rich regions were extensively surveyed during the year (*Mining Magazine*, May 1991).

The non-ferrous metals industry in China was designated as a priority industry for expansion during the years 1986 to 1990. China produced 2.35m. tonnes of ten non-ferrous metals in 1990—copper, aluminium, lead, zinc, nickel, tin, antimony, mercury, magnesium and titanium—an increase of 7.2 per cent over the previous year. China is now the world's fourth largest non-ferrous metals producer, following the United States, the former Soviet Union, and Canada. During the Seventh Five-Year Plan period, although China's non-ferrous metals industry experienced a rapid development, the supply of non-ferrous metals still fell short of demand and therefore China had to import large quantities of the metals. During the 1978–82 period, China imported 1.67m. tonnes of non-ferrous metals and exported 372,200 tonnes (*Beijing Review*, 4–10 Mar. 1991). The situation has changed since 1987, but the non-ferrous metals industry is still a weak link in the national economy.

Before the 1980s, China's non-ferrous metals were produced under a highly centralized mandatory plan. The state drew up production plans and was responsible for the supply of raw materials, and for the allocation of products, prices, and construction funds. The enterprises were not responsible for their profits or their losses. From 1979, however, China adopted a series of measures to eliminate defects in its economic structure and improve heavy industry. All these brought about some important changes in non-ferrous metals production and the methods of operation during the Seventh Five-Year Plan period.

China has, for example, minimized the scope of mandatory plans and expanded the scope of guidance plans as part of its reform of the state planning structure. The varieties of non-ferrous metals under the mandatory plans decreased from more than 80 to 23 in 1985 and to 17 in 1988 (*Beijing Review*, 4–10 Mar. 1991).

Antimony, mercury, magnesium, titanium, rare metals, rare earth, and other metals are now available on the market. Even those produced under the state mandatory plans can be sold at a market price. Statistics show that the proportion of copper, aluminium, lead, zinc, and tin sold at market prices increased from 26.7 per cent in 1983 to 63 per cent in 1990.

The Chinese steel industry is faced with the problem of meeting strong demand through quantitative expansion, while simultaneously improving quality. The Seventh Five-Year Plan promoted substantial renovation of existing facilities in order to boost production capacity. Three-quarters of China's iron and steel industry's investment during the period was used to re-equip old enterprises. Financing for the expansion was from domestic as well as foreign sources. Between 1986 and 1990, China added 5.12m. tonnes of steel production capacity by way of increased government funding and raw materials allocation. Rolled steel output during the period rose 27 per cent, reaching 51.5m. tonnes in 1990. Despite such an increase in output, China remains fourth in world steel production after the former Soviet Union, Japan, and the United States.

Much of the industrial emphasis in the late 1980s was on improvement of quality control and enhanced productivity. China's iron and steel industry has made further progress in the quality and variety of its products. Sixteen varieties of steel fitting international standards were produced during the Seventh Five-Year Plan period. By 1990, more than 40 per cent of China's iron and steel enterprises were operating according to international norms.

Another priority of the industry was to increase savings of energy during steel production by China's primary iron and steel plants. During the Sixth Five-Year Plan period, China's rolled steel output increased by nearly 38 per cent, whereas energy consumption in the steel sector grew by only 7.3 per cent, resulting in a saving of 10m. tonnes of standard coal. By the end of 1990, China's coal consumption for each tonne of steel output dropped to 1.63 tonnes, saving a total of 9m. tonnes more than in the Sixth Five-Year Plan period (*Beijing Review*, 24–30 June 1991).

Although China's present development strategies encourage accelerated prospecting for energy resources in an effort to reduce the nation's unbalanced reliance on petroleum, increased prospecting is not the major priority. Current policies for petroleum, for example, focus more on transforming existing enterprises than on accelerating exploration for new deposits. New efforts have been made to renovate and expand old oil fields and refineries, and provide them with supporting facilities. In attempts to maintain a stable oil yield at Daqing, a great deal of equipment has been imported from abroad, and advanced technology has also been introduced on a large scale. Since petroleum is a major foreign exchange earner, the Chinese government realizes it must continue to exploit oil resources and at least maintain current levels of production while stepping up exploration efforts. Even though China is a major world oil producer, it exports nearly 25 per cent of its production in return for as much as one-quarter of its foreign exchange earnings.

In 1990, total production of crude oil reached 138m. tonnes—an increase of just 8m. tonnes over 1985—far short of the anticipated 25m. tonnes increase. China's Seventh Five-Year Plan had called for production to reach 150m. tonnes by 1990. A primary factor behind the slow-down in China's petroleum sector was the stagnation of onshore production after 1985. Output at all major eastern fields such as Daqing, Shengli, Huabei, and Liaohe has levelled off or begun to decline, more than offsetting continued gains from onshore fields such as Jilin and Jiangsu and from offshore regions. Production from the north-west has also continued to increase, reaching 136,000 barrels per day (b/d) in 1990, including 18,000 b/d from the Tarim Basin. Total crude production rose only 0.45 per cent in 1989, matched by a 0.46 per cent increase in 1990. In 1991, crude production rebounded only slightly, rising 0.91 per cent over 1990 to 139.57m. tonnes.

Owing to China's increasing need for energy supplies, Chinese geologists accelerated their prospecting for natural gas in the 1980s. No detailed assessment of China's natural gas reserves has yet been completed, but experts agree that today's known reserves are likely to be a small fraction of the full potential.[8] Sichuan province is presently China's largest natural gas producer, accounting for more than 40 per cent of the country's output. Prospecting for onshore natural gas is now under way in many parts of China, including the eastern coastal plain, where shallow gas deposits exist. Associated gas reserves of about 45 mtoe (million tonnes of oil equivalent) may be present in the Zhongyuan Basin (Hayes 1986), which lies between Henan and Shandong provinces. The Zhongyuan Basin is expected to soon replace Daqing (associated gas) as eastern China's largest gas producer. During the 1986–90 period, natural gas extracting capacity increased by 1,538m. cubic metres, meeting the plan targets. Production in 1990 reached 15,298m. cubic metres, or 13.8m. tonnes oil equivalent, just 1.65 per cent over 1989 levels, but more than 11 per cent over 1985 quantities.

For coal, like petroleum, emphasis during the 1986–90 period was not on the exploration for additional resources, but rather on the technological transformation and renovation of the coal industry. Specifically, the Seventh Five-Year Plan called for (1) improved efficiency of coal-mining enterprises; (2) technological transformation, renovation, and expansion of existing coal-mines, (3) the pooling of funds to operate existing mines, and (4) the provision of aid to the construction of selected small and medium-sized local coal-mines (Xinhua News Agency News Bulletin 1986: 20–1). At the beginning of the plan period the former Ministry of Coal Industry announced China's ambitions to modernize 77 major coal-mines before 1990 (Xinhua News Service, 18 July 1986). These mines were scheduled to serve as models in productivity, with each worker producing 2 tonnes per work day, an amount which is more than double the present figure for the nation. While tunnelling, extraction, loading, and conveying are more than 95 per cent mechanized in most Western nations, the degree of mechanization in China, even in the country's largest and newest mines, is less than 50 per cent (Economist Intelligence Unit 1990).

China's Seventh Five-Year Plan output target of 1bn. tonnes of coal was met and exceeded in 1989, when 1.054bn. tonnes were produced. Coal production capacity grew by 130m. tonnes during the entire five-year period. The 1980s experienced rapid development of the Chinese coal industry, characterized by large-scale investment in centrally controlled mines, promotion of local mines, and expansion of collective and individually operated (village or *xiangzhen meikuang*) mines. Although the coal industry ended the decade with output two-thirds higher than in 1980, much of the growth came from local mines, or state-owned mines under control of the province, county, or district, which accounted for 75 per cent of the increase between 1980 and 1988; collective and individually run mines in turn accounted for 70 per cent of the increase in local mine output (Fridley 1991*d*).

Coal remains the most important source of primary energy in China, providing over 70 per cent of the country's total needs. Government leaders have therefore given coal the highest priority in their overall energy and industrial development strategies. Output is still, however, projected to lag behind domestic demand by 120m. tonnes by the end of the century. Foreign investment was actively sought to develop the sector during the Seventh Five-Year Plan period; for example, an agreement for the planning and design of coal projects was signed with the United States company Bechtel, and a joint-venture project involving the expenditure of $725m. was in operation at the giant Antaibao opencast coal-mine in China's northern Shanxi province. Most of the mine's production facilities were completed in 1990, three years after its official opening.

Though the Seventh Five-Year Plan period was still incomplete, changes occurring in China's economy late in the period led government officials to adjust economic strategies beginning in 1988. A Three-Year Austerity Programme (1988–91) was introduced in an effort to cool the nation's overheated economy. Faced with soaring inflation, corruption, and a general economic slowdown, China's market-oriented Seventh Five-Year Plan policy was questioned by some Chinese government officials. Civil unrest, violence, and political turmoil in 1989 extended the doubt among China's leaders about the appropriateness of the market incentives incorporated into the original Seventh Five-Year Plan. In a change of direction, price controls were reimposed on key raw materials in 1988, signalling failures in China's ambitious reform programme to phase out central planning and allow the market-place to set prices. The Chinese government also ordered hundreds of trading companies to halt business and imposed tighter controls on the bank accounts of thousands more in a growing anti-corruption campaign.

THE EIGHTH FIVE-YEAR PLAN (1991–1995)

After prolonged debate, China's Eighth Five-Year Plan for National Economic and Social Development (1991–5) was approved on 9 April 1991 by the Fourth Session of the Seventh National People's Congress. At the meeting, the Outline of the Ten-Year Programme (1991–2000) to chart the course for the nation's construction and development in the present decade was also adopted.

China's newest five-year plan has the difficult task of restoring economic order while deepening reform. The Eighth Five-Year Plan seeks to reverse the environment that persisted throughout most of the 1980s—when industry grew much faster than other sectors, leading to inflation and an overheated economy. Conditions will be created to facilitate sustained, balanced economic growth. The emphasis will be placed on the readjustment of China's national economic structure, the improvement of economic returns and upgrading of existing enterprises rather than initiating new ones, furthering the reform and opening to the outside world, and diminishing the role, scope, and importance of direct government planning while promoting market forces.

The target for average annual economic growth was set at 6 per cent between 1991 and 1995. This is less than the average growth rate in the 1980s, but sufficient for China to meet and exceed its goal of achieving output in the year 2000 equal to four times the gross national output of 1980. The 1995 gross national product (GNP) will reach 2,325bn. yuan (calculated in 1990 prices), a 33.6 per cent increase over 1990 (*Beijing Review*, 13–19 May 1991).[9] The total agricultural output value will rise to 878bn. yuan, an 18.9 per cent increase over 1990 or an average annual increase of 3.5 per cent, and the industrial output value will reach 3,270bn. yuan, 37.1 per cent more than in 1990 or an average annual increase of 6.5 per cent.

The decline of China's energy industry during the latter part of the Seventh Five-Year Plan period prompted government planners to designate the sector a higher priority than is usual in the nation's five-year plans. Development and expansion of the industry will coincide with an emphasis on energy savings. In 1995, the national output of energy will be up to 1.172bn. tonnes of standard coal, an increase of 132m. tonnes over 1990 production. Over the five-year period to 1995, a total of 100m. tonnes of standard coal will be saved through conservation and efficiency gains. Energy consumed for every 10,000 yuan of GNP will be reduced from 9.3 tonnes of standard coal in 1990 to 8.5 tonnes in 1995, an average annual energy conservation of 2.2 per cent.

Coal production is scheduled to increase to 1.23bn. tonnes by 1995, 150m. tonnes more than in 1990. Construction will continue at the Huolinhe, Yimin, Yuanbaoshan, and Jungar opencast coal-mines in Inner Mongolia, the Datong coal-mining area in Shanxi province, the Shenfu–Dongsheng mining area of Shaanxi and the Tiefa and Shuangyashan coal-mining areas in north-east China, and the Yanzhou, Huainan and Yongcheng coal-mining areas in east and south-central China. The construction of new projects will also start in the Huangling coal-mining area in Shaanxi, the Lingwu coal-mining area in Ningxia, and the Anjialing opencast mine at Pingshuo in Shanxi province. Emphasis will be given to the mechanization of production, the expansion of large opencast mines, and the building of new ones.

As in the previous five-year plan, the objective for development of China's oil industry will be to stabilize production in traditional east China fields while developing new ones in west China. The output of crude oil in 1995 will top 145m. tonnes (including 5m. tonnes from offshore fields), up 7m. tonnes from 1990 production. Onshore production will increase just 2m. tonnes per year between 1991 and 1995. Exploration and development activities are to be expanded at the long-established Daqing, Shengli, and Liaohe major oil fields to help ensure stable increases in crude petroleum output in east China. Increasing efforts are having to be made at maintaining current levels of production at Daqing—about 55m. tonnes a year—or nearly 40 per cent of the country's total output (*Petroleum Economist* 1991).

World crude steel output in 1991 was 13.8 per cent down from record levels the previous year; however, Chinese production increased by 6.9 per cent more than in 1990 to 70.6m. tonnes in 1991. This increase in output was in line with China's official policy to boost annual crude steel capacity during the 1990s from current levels to 72m. tonnes by 1995, about 6m. tonnes more than in 1990. China will far surpass this target, as steel output in 1992 reportedly approached 80m. tonnes, an increase of nearly 13 per cent over the previous year (*Mining Journal*, 5 Feb. 1993). Since the start of major reforms in the industry 14 years ago, the number of plants producing over 1 million tonnes per year (mt/y) of steel has risen from 12 to 17, while those producing over 500,000 tonnes per year (t/y) had increased from two to 21.

Between 1991 and 1995 the emphasis will be on continued modernization of China's key iron and steel industrial bases by improving production, adopting Western-style management techniques, and introducing new computer hardware and software to boost efficiency. According to the Eighth Five-Year Plan, the output of rolled steel will reach 59.5m. tonnes in 1995; as with crude steel, this target is likely to be exceeded given the level of investments planned for the steel industry. By 1995 rolled steel produced at the level of international norms will be 50 per cent, compared to 40 per cent in 1990, steel plates and pipes 40 per cent against 37 per cent, and continuously cast steel 35 per cent compared to 22.4 per cent (*Beijing Review*, 13–19 May 1991). Product quality will continue to be improved and diversified according to the Plan, and energy savings increased. Consumption of energy for every tonne of steel will decrease from 1.63 tonnes to 1.55 tonnes of standard coal.

Like iron and steel, China's non-ferrous metals industry will continue to undergo reform and structural change during the 1990s. Management and decision-making will become less centralized, and production targets will be increasingly set by industrial enterprises based on profit/loss margins and economic efficiencies. The reform of the commodity pricing system will continue, allowing for two sets of prices for non-ferrous metals, a planned price and a market price. The chronic shortage of non-ferrous metals caused a rapid rise in the market price during the Seventh Five-Year Plan period, especially in 1988 when China's inflation rate was soaring owing to an overheated economy. In the 1990s, China's non-ferrous metals industry will be more fully integrated with the world market, as more and more foreign trade companies, special economic zones, and coastal open areas have been established. While China will rely on its own resources and technology to expand its non-ferrous sector, it will continue to import advanced foreign technology and equipment.

According to the China Nonferrous Metals Industry Corporation (CNNC), the output of ten primary non-ferrous metals will increase by 5 per cent annually to reach 3.5m. tonnes in 1995. Much of the new production capacity will be based on technical renovation of existing mining complexes. During the Seventh Five-Year Plan, 9bn. yuan was set aside for the technical renovation of non-ferrous metal enterprises, 60 per cent of which was allocated by the enterprises themselves. Government reports indicate that of the added production capacity between 1986 and 1990, two-thirds was attributed to technical reform.

As when the Seventh Five-Year Plan was drawn up, the initial Eighth Five-Year Plan was established during a period of relatively slow economic growth; thus it focuses on guiding the economy toward resumed, stable growth. Together with the Eighth Five-Year Plan, the Outline of the Ten-Year Development Programme (1991–2000) was adopted to strengthen the effectiveness of the Eighth Five-Year Plan and to create general goals for guiding China's economy through the 1990s and into the next century. The Ten-Year Development Programme has five main objectives: (1) to quadruple China's 1980 GVAIO by the end of the century; (2) to raise living standards; (3) to further develop education, promote progress in science and technology, improve economic management, and readjust China's economic framework; (4) to begin to create an economy based on public ownership that combines a planned economy with market regulation; and (5) to strengthen the legal system and culture (*Beijing Review*, 15–21 Apr. 1991). Output targets to 2000 established by the ten-year strategy include 1.4bn. tonnes for coal, 80m. tonnes for steel, and 120m. tonnes for chemical fertilizers.

A GROWING RESOURCE CRISIS

Nearly a decade and a half of China's open-door policy has demolished the foundation of Chairman Mao's development strategy—resource self-sufficiency. During the 1980s, China's reliance on imported supplies of oil and minerals rose dramatically to satisfy pent-up demand and meet the needs of an overheated economy. When the three-year economic readjustment programme was initiated in October 1988, economic growth in China slowed from double-digit rates to only 3.9 per cent growth in gross domestic product (GDP) in 1989. The slow-down extended into 1990, and industrial output stagnated. The downturn did for the time being ease one of the most crucial problems resulting from a decade of rapid growth—a growing shortage of energy and mineral resources in what has historically been one of the world's largest producers of coal, oil, steel, and other primary industrial commodities.

The deteriorating state of China's energy and mineral industries today is reflected in the slow-down in output growth of these sectors, especially since 1988. Despite the stimulus of price reform in 1983, continued controls on output prices and allocation have left many of the energy and mineral industries awash in red ink as high inflation eroded their largely fixed revenues. Profits of 11bn. yuan ($3.74bn.) in 1985 turned into a 2.9bn. yuan deficit by 1989. Similarly, state subsidies to the loss-making coal industry continued to grow, while the persistent financial shortfall in the natural gas sector forced producers to wrest increased output from existing wells without adequate investment in exploration and new production facilities.

Since the early 1980s China's consumption of metals and non-metallic mineral resources grew dramatically with the need for more raw materials in industry and the military. Rates of growth of metals consumption in China between 1981 and 1991 were between twice and 15 times those of the world for aluminium, cadmium, copper, lead, magnesium, nickel, steel, tin, and zinc. Consumption of steel, the nation's most costly imported product, rose at an average annual rate of more than 6 per cent during the 1980s. The potential for future consumption growth is clearly huge for many metals and, even at current high levels of production, China's domestic output of some metals is falling far short of demand.

Supply has indeed lagged, forcing China to fill the gap through costly imports. By 1989, for example, China had become a major importer of iron and steel, manufactured fertilizer and non-ferrous metals. Despite an extensive minerals base, constraints on hard currency increasingly meant that the Chinese government had to delay capital investments in mine and plant construction and plant expansion.

Remarkably, even at a slower-than-recent GDP growth rate of around 3.4 per cent a year through the 1990s, it is expected that China's consumption of steel will rise by nearly 5 per cent annually, reaching as much as 148m. tonnes in 2000; output is projected to be around 85m. tonnes. The shortfall of over 60m. tonnes could be met through imports, but at a cost of at least $10bn.

China's domestic supply of other valuable metal commodities, such as copper, nickel, and aluminium, will continue to fall short of demand in this decade. The situation for energy is even worse. Until the oil price crash of 1986, China exported up to 80 per cent of its increased oil production each year, squeezing the domestic market in favour of easy export income. With the price of its benchmark crude, Daqing, falling to as little as $10 per barrel that year, China quickly reversed its policy, allocating greater volumes of oil to domestic refiners and allowing exports of crude and petroleum products to slide.

Nevertheless, demand for some products, particularly diesel fuels, has continued to outpace supply. In 1986, China reversed its long-standing policy against petroleum product imports and began a sustained programme which reached 156,000 barrels per day (b/d) of imports in 1992. With the slow-down in domestic oil production, supplies of crude oil also tightened, and since 1988 imports soared from near zero to more than 225,000 b/d. Exports of oil continued to fall from 1988 to 1992, and low domestic oil prices have only recently been deregulated. China's oil demand is expected to be more than 200m. tonnes in 2000.

The decline in China's energy and mineral industries will create near-intractable problems for government planners in the 1990s. Already, top energy officials have admitted that China faces a long-term chronic shortage of energy. In 1979, a long-term energy plan was devised to facilitate the quadrupling of China's gross industrial and agricultural output by the year 2000. By assuming a primary energy elasticity of demand of no less than 0.5 and no less than 1 for electric power, the quadrupling of economic activity could be supported by a doubling of primary energy production and a quadrupling of electric power output (Fridley 1991*d*). Additionally, an implicit assumption was that China could remain energy self-sufficient; hence, domestic production targets were set to achieve this long-range goal. Based on 1989 output levels, the targets suggest an annual average rate of growth for the remainder of this decade of 3.5 per cent for crude oil, 2.9 per cent for coal, 6.8 per cent for hydropower, and 6.6 per cent for natural gas, or 3.3 per cent for total energy production.

Like energy, mineral shortages are also foreseen in the coming years, prompting the Ministry of Geology and Mineral Resources in 1989 to request a 40 per cent increase over its present share of the national budget to boost surveying activities. Shortages of steel, aluminium, and fertilizers will continue through the 1990s, and meeting the predicted shortfall by imports could cost up to $20bn. by 2000. Imports of energy and minerals could eventually total $40–50bn., nearly equivalent to China's total export revenues in 1990. With expected strong growth in the import of other commodities such as grain and capital equipment, China is unlikely to be able to satisfy growth in energy and minerals demand. But without sufficient external supplies of these commodities, China may face bottlenecks in the production of primary goods, the effects of which will ripple throughout the economy.

CONCLUSIONS: PLANNING AND DEVELOPMENT

The mineral-related industries of China are largely affected by the nation's planning system. As has been demonstrated, the Chinese national economy is managed and administered according to specific regulations to stimulate economic growth and prosperity. Minerals and energy planning is a crucial component of China's five-year development plans, economic reforms, and annual plans. For practical purposes, annual plans are the most important in China as economic and industrial policies are periodically adjusted to match the socio-economic conditions of the nation. Several issues dominate the Chinese planning system, including the sectoral distribution of investment funds, ideological and political concerns, and the division of power among central government, provincial, and local authorities.

Mineral exploration, development, and trade activities are all entered into Chinese development strategies. Mineral activities during the First Five-Year Plan period (1953–7) stressed industrialization and large-scale increases in production capacities. The Second Five-Year Plan period (1958–62) was interrupted by the Great Leap Forward, which aimed to accelerate economic growth in China, but instead created havoc in the economy. The sudden withdrawal of Soviet technicians in the early 1960s further perturbed all industrial activities, including those related to minerals exploration and development. By the Fourth Five-Year Plan (1971–5) the Chinese minerals economy had recovered from the Cultural Revolution and earlier interruptions, and prospered through the end of the decade albeit with continued cycles of growth and slow-downs. The Sixth and Seventh Five-Year Plans (1981–5 and 1986–90) are noted for their introduction of market reforms and the 'open-door' policies, which permit foreign economic exchanges with China in the hope that they will facilitate modernization and economic development. The Eighth Five-Year Plan (1991–5) will facilitate greater market influence within a planned economy framework. In the 1980s and 1990s China has invited foreign participation in the form of joint ventures to accelerate the development of its mining sector.

Except for disturbances caused by the Great Leap Forward in the late 1950s, the withdrawal of Soviet technicians in the early 1960s, and the Cultural Revolution in the late 1960s, the Chinese mining industry advanced at a relatively smooth and steady pace, particularly in terms of output. When production targets were established in accordance to resource limitations, expected capital flows, and existing technologies, they were generally met or exceeded. However, since the achievement of ambitious output targets received the greatest emphasis in government plans, little consideration was given to bettering efficiencies in resource utilization, conserving resources, and improving safety standards at the nation's mines.

NOTES

1. Zhou Enlai divided the growth rates of industrial output into three categories: 'leap forward', 'great leap forward', and 'exceptionally great leap forward'. An annual increase from 20 to 25% is a leap forward; above 25% and below 30% is a great leap forward; and 30% or above is an exceptionally great leap forward. For details on his definitions, see his 'Report on the Adjustment of Major Targets in the National Economic Plan for 1959

and Further Development of the Campaign for Maintaining Production and Practicing Economy', printed in *New China Semi-Monthly*, Peking (Zho 1959).

2. The 'general line' for socialist construction was simply a set of four party slogans—abundance, speed, better, and thrift.
3. Note that Chinese production figures for 1958, 1959, and 1960 are suspect because it is believed that, where assigned quotas could not be met, local Chinese communist authorities may have falsified production figures.
4. Red Guards, technicians, peasants, and coal-mine workers were among the demonstrators.
5. Zhou Enlai first mentioned his 'four modernizations' concept at the Third National People's Congress in 1964. The concept was indeed supported by the Chinese Communist Party for many years. In early Jan. 1975 Zhou unveiled his four modernizations programme at the Fourth National People's Congress, which overwhelmingly stressed the need for unity. Zhou introduced his four modernizations programme in two stages: first, to build 'an independent and relatively comprehensive industrial and economic system' by 1980; and second, to 'accomplish the comprehensive modernization of agriculture, industry, national defence, and science and technology before the end of the century' so that the Chinese economy will be on the level of other developed nations (Riskin 1987).
6. Central Committee of the Communist Party of China (1984).
7. To reach this target, the plan envisaged two major stages: (1) the laying of a solid foundation during the 1980s, and (2) a period of vigorous economic expansion in which the growth rate should exceed 8% or more p.a. during the 1990s. The tasks in the first stage are to strengthen agriculture, energy resources, transport, and communications, to reorganize industries and increase labour efficiency, and to raise the levels of science, technology, and education. Only on such a solid foundation could China hope to achieve the rapid rate of growth expected in the second stage. An important component of these plans is reduction of the population growth rate to keep China's population within 1,200m. by the turn of the century.
8. Reserve estimates of natural gas in China have ranged from 120 mtoe by the World Bank, which would last under 10 years at current rates of consumption, to an optimistic Chinese forecast of 6,750 mtoe. If the latter estimate were true China's natural gas reserves would be third largest in the world after the former Soviet Union and Iran and sufficient to last more than 450 years at current consumption rates.
9. In a move to conform to international standards, China's State Statistical Bureau began tabulating GNP figures in the late 1980s. The Bureau defines GNP as the sum of value-added for all sectors—'material production' and 'non-material production'—of the economy, and depreciation of fixed assets. For figures after 1980, the difference between the sum value of primary, secondary, and tertiary industries and GNP is net foreign factor income.

PART II

ECONOMIC RELATIONS AND IMPACTS

4
The Mining Base and Industry of China

THE MINERALS AND ENERGY ENDOWMENT OF CHINA

With its complex geology and huge land area (China ranks second after the Russian Federation in area), China is well endowed in most major metallic, non-metallic and mineral fuel resources. Despite the vast minerals and energy base, however, with a population of more than 1.1 billion people in China, the minerals and energy position per capita is below that of most other resource-rich countries. In addition, although the nation's endowment of most non-fuel minerals is large, many of the deposits are of poor quality or in remote locations,[1] making them very costly and sometimes difficult to exploit.

It was not until the end of 1985 that the Chinese government undertook the first detailed and comprehensive nationwide survey of mineral and energy resources and mines in the country.[2] According to that survey and others since,[3] geologists in China have discovered more than 160 types of mineral and energy resources and have verified reserves of most types. These include five types of ferrous minerals, 20 types of non-ferrous and precious metallic minerals, 76 types of non-metallic and geothermal and ground water minerals, and six types of mineral fuels (Gao 1986). Mineral ores in China are distributed in over 200,000 locations (Ministry of Geology and Mineral Resources 1991*b*), although the distribution is fairly uneven throughout the country due to the complicated and varied geological conditions, and there are different degrees of intensity of geological investigation in China.

Reserves of many types of mineral and energy resources in China are the largest in the world (Table 4.1),[4] and several minerals, including rare earths and tungsten, occur in an abundance that not only satisfies domestic demand, but also allows for exports. Some of the few minerals where China's reserves can be considered relatively small include chromite, platinum-group metals, titanium (rutile), and zirconium.

With its production level reaching 2.52m. tonnes in 1991, China became the world's fourth largest non-ferrous metal producing country in the world, behind the former Soviet Union, the United States, and Canada. China's Eighth Five-Year Plan calls for the industry to grow at an average annual rate of 5 per cent in the 1991–5 period in line with national economic development. China's major non-ferrous minerals (including precious metals) include copper, aluminium, lead, zinc, nickel, cobalt, tungsten, tin, molybdenum, mercury, antimony, bismuth, gold, silver, rare earths, and the platinum-group metals. Of these, exploitable

TABLE 4.1. *Estimated mineral and energy reserves[a]/resources of China*

Mineral	Reserves: Amount ('000 tonnes, unless otherwise noted)	Share of world (%)	Potential
METALS			
Antimony	2,180[b]	51.9	World's largest
Bauxite	150,000	0.7	Sizeable, low grade
Cadmium	15[c]	2.8	Moderate
Gold	0.4[d]	NA	Moderate
Iron ore	3,500,000	5.4	Sizeable, low grade
Lead	6,000	8.5	Moderate
Manganese	14,000	1.7	Moderate
Mercury	300[c]	8.1	Sizeable
Molybdenum	500[c]	9.1	Sizeable
Nickel	725[b]	1.5	Poor
Rare earth metals	43,000[f]	51.3	World's largest
Tin	1,500[c]	25.4	World's largest
Titanium	30,000[b]	14.7	Sizeable
Tungsten	1,050[b]	45.7	World's largest
Vanadium	600[b]	14.2	Sizeable
Zinc	5,000	3.3	Moderate
NON-METALS			
Baryte	40,000	24.0	World's largest
Fluorspar	27,000[g]	11.3	Sizeable
Industrial diamonds	10,000[h]	1.0	Poor
Phosphate	210,000	1.6	Sizeable, low grade
Talc	NA	NA	Sizeable
ENERGY MINERALS			
Coal	114,500,000[i,j]	11.0	Sizeable
Natural gas	1.0[k]	0.8	Moderate
Petroleum	3,200[l,m]	2.4	Sizeable
Uranium	50	1.7	Moderate

Notes: NA not available. Figures are rounded.

[a] Reserves in China are typically categorized as 'industrial' and 'non-industrial' or 'prospective'. Industrial reserves are technologically recoverable or mineable, with some consideration of economics. Non-industrial reserves are not technologically recoverable for industry (see n. 4, this chapter). Reserve figures in this table are primarily from Western sources, which are generally based on publicly released Chinese data. For consistency reasons it was deemed important to list either Western or Chinese reserve estimates, but not both. Western reserves are defined as recoverable material in the reserve base that can be economically extracted or produced at the time of determination.

[b] Contained metal.

[c] Metal.

[d] Gold bullion reserves.

[c] '000 76 lb. flasks.

[f] '000 tonnes of rare earth oxides (REO).
[g] Contained fluorspar.
[h] '000 carats.
[i] Proved reserves of coal are generally taken to be those quantities which geological and engineering information indicate with reasonable certainty can be recovered in the future from known deposits under existing economic and operating conditions.
[j] Includes anthracite, bituminous, sub-bituminous, and lignite.
[k] Trillion cubic metres.
[l] Million tonnes.
[m] Speculative reserve estimates due to lack of detailed exploration and estimation of western onshore basins.

Sources: US Bureau of Mines (1982–93 (1993)); Crowson (1986–92 (1992)); Geological Publishing House (1992, 1990); British Petroleum Company (1984–92 (June 1992)); *Mining Journal* (24 Mar. 1989: 237).

reserves of tungsten, antimony, tin, and molybdenum are among the largest of any country in the world. About 41 per cent of the world's estimated tungsten resources are in China (US Bureau of Mines 1982–93 (1993): 187), as well as a majority of the world's known reserves. Tungsten-producing areas in China are mainly concentrated in southern Jiangxi, south-eastern Hunan, and northern Guangdong. There are presently nearly 100 tungsten mines in China, including the world's three largest mines in Jiangxi, Hunan, and Guangdong provinces.

Gold and silver, two precious metals, are receiving high priority in terms of exploration and development activities in China during the Eighth Five-Year Plan. China will continue large-scale mining activities in order to boost gold output and create more efficient gold-mining businesses. The National Gold Administrative Bureau has instituted temporary regulations for large-scale enterprises and set production capacity levels for 30 gold projects undergoing technical renovation in the period 1991–5 (*Mining Journal*, 6 Mar. 1992). Small and medium-sized gold-mines, which contribute nearly 80 per cent of China's total gold output, are hampered by low rates of efficiency and utilization, and extraordinarily high mining costs.

In 1992 China expected its gold industry to expand in size by more than one-tenth over the previous year. Government reports indicated that investments totalling 7–8bn. yuan ($1.3–1.5bn.) would increase gold output during 1991–5 by around 40 per cent compared with the preceding five-year period (*Mining Journal*, 6 Mar. 1992). Investment totals will be nearly double the amount invested under the 1986–90 plan. In 1992 alone, capital investment into the gold industry was budgeted at 1.3bn. yuan, of which more than three-quarters was to be spent on capital construction projects and renovation. China's gold development activities are focusing on construction of 11 key gold projects including seven new mines in

Liaoning, Sichuan, Shandong, Henan, and Shaanxi provinces and expansion projects at mines in Jilin and Hubei provinces and the Inner Mongolia autonomous region.

China has become a major gold producer, ranking sixth worldwide in 1991, and its gold reserves are among the largest in the world. Gold output increased at an annual rate of 11 per cent over the past decade, and is forecast to drop to an annual rate of 9 per cent during the Eighth Five-Year Plan (1991–5). The number of geological workers engaged in gold exploration increased from 52,000 in 1985 to more than 90,000 by the end of the decade. Nearly 1,000 county-level gold mines exist throughout China, and more than 4,000 gold deposits have been discovered throughout the nation (Wang Xionglin 1988: 114). China's main gold production centres are in Shandong, Henan, Heilongjiang, Shaanxi, Gansu, Sichuan, and Hebei provinces.

Gold is widely distributed in China and occurs mainly as vein-type gold in mountainous regions (about 50 per cent), placer gold (about 10 per cent), and other types of ore deposits where gold is recoverable as a by-product. Vein-type gold constitutes approximately half of the proven reserves. Two primary gold production bases emerging in China—one between south-west Guizhou province, north-west Guangxi Zhuang autonomous region, and south-east Yunnan province, and the other between Sichuan, Gansu, and Shaanxi provinces. Gold-mine belts are also found along the middle and lower reaches of the Yangtze River, mountainous regions (Ailaoshan) in Yunnan, and in the northern part of the Junggar Basin in Xinjiang Uygur Autonomous Region. New exploration techniques have aided the search for gold in Henan, Zhejiang, and Heilongjiang provinces. Shandong leads the country in both reserves and annual gold production.

An important group of non-ferrous mineral resources in China are the rare earth metals. According to Chinese sources (Wang 1988; Gao 1986; Hu, Zhang, and Liu 1984) and the US Bureau of Mines (1982–93 (1993)), the nation possesses the largest reserves of industrially valuable rare earth minerals (estimated at 43m. tonnes) in the world. A large portion of China's niobium, tantalum, and dispersed elements occur in iron and other metallic or non-metallic deposits. The Bayan Obo Iron Mine in Inner Mongolia contains over ten rare earth elements and is confirmed to be the largest rare earth metals mine in the world. Bayan Obo accounts for nearly 90 per cent of China's proven rare earth reserves (Hu, Zhang, and Liu 1984: 95),[5] and it has a full range of rare earth elements of high quality. Other rare earth deposits have been discovered in Anhui, Gansu, Guangdong, Guangxi, Henan, Hunan, Jiangxi, Shandong, Sichuan, and Hebei (US–China Business Council 1988), with many of the rare earth deposits in southern China being associated with weathered sediments.

Several ferrous metallic mineral resources occur in abundance in China, including iron ore, manganese, chromite, vanadium, and titanium. With the exception of chromite, the identified reserves of all of these minerals are large. Despite the apparent abundance of iron ore, however, the Chinese government is resigned to importing the metal well into the next century, as the increasing demand for iron ore for industrialization will greatly exceed supply. Moreover, though China's iron ore resources are large, they are mostly low-grade, averaging less than 35 per cent iron. Imports of high-grade iron ore are therefore needed to satisfy domestic demand.

According to the Ministry of Geology and Mineral Resources, there are approximately 24.6bn. tonnes of economically recoverable iron ore reserves in China. Proven iron ore deposits in the nation are widely scattered in China but most concentrated in a few areas. China's iron ore deposits in the north-east, north, and south-west make up 71 per cent of the national total (Lu Zhen 1988). Chinese iron ores are characterized by their relatively low quality as well as their association with other metals, for example vanadium and titanium and, to a lesser extent, chromite, copper, cobalt, nickel, gallium, and scandium.

One of the few minerals China lacks in abundance is chromite. Chromite deposits are principally located in Xinjiang, Tibet, Gansu, and Qinghai, with lesser amounts existing in Jilin, Heilongjiang, Inner Mongolia, Yunnan, Jiangsu and other parts of the nation. Xinjiang produces the largest share of chromium metal in China, and may be the only exporter of chromite ores in the country. At present all of the chromite ores in Xinjiang are extracted from the Junggar Basin, with the largest share from the Salthohai mine in the basin's north-western region.

As for non-metallic resources, a total of 76 non-metallics have identified reserves in China, including several metallurgical and chemical raw materials. China is a large producer of a variety of industrial minerals, with significant exports of barite, fluorspar, salt, talc, and cement. However, output levels of some industrial minerals are insufficient to satisfy domestic requirements, necessitating imports of several commodities including soda ash, potash, and other fertilizers. Phosphate, an important fertilizer raw material, is abundant in China, though in general of low quality. Most of the large phosphate deposits of China are distributed in two main belts in the southern part of the country: the Hubei–Hunan–Guizhou belt of Sinian age and the Sichuan–Yunnan–Guizhou belt of Cambrian age. Because identified reserves of phosphate in northern China are not large and are of relatively low grade, some phosphate has to be transported from the south to the north. The most important phosphate-producing districts in China are Kaiyang of Guizhou, Kunyang of Yunnan, and Xiangyang of Hubei.

Another non-metallic mineral resource abundant in China is gypsum, which is extensively distributed in 19 provinces and autonomous regions in the nation. According to Chinese sources, China is endowed with the largest gypsum reserves in the world. Hubei, Xizang (Tibet), Shandong, Jiangsu, Ningxia, Shanxi, Shaanxi, Hunan, Gansu, and Anhui contain the richest gypsum deposits in China. Of these provinces and autonomous regions Hubei is the largest producer, and is noted for its 400-year-old mine at Yingcheng. Most of China's gypsum production is consumed by the nation's cement-manufacturing industry.

Owing to the complexity of China's geology and vastness of the nation's territory, precious stones, including diamonds, sapphires, rubies, topaz, gems, and jade, have been found throughout China. More than 170 types of gems and jade ores have been discovered in the country, with nearly half of them being of high quality (*China Daily*, 14 June 1993). During the past several years China reportedly uncovered more than 200 deposits of gems and jade. Provinces and autonomous regions well endowed in gem and jade resources include Xinjiang, Yunnan, Liaoning, Shandong, and Henan.

China currently produces more than 2,000 tonnes of jade annually, by the more than 1,500 state-run or collective mills, in addition to greater than 10,000 individual processors. Trade in Chinese gems and jade has been developing rapidly during the 1990s, with the yearly export value of jewellery and jade handicrafts now exceeding $69m.

Reserve amounts of diamonds in China are not clear, but geological conditions in the nation seem favourable for diamond occurrences. In 1990 the Xinhua News Agency announced that a diamond weighing 60.5 carats with an axial length of 22.5 mm was unearthed from the Wafangdian mine in Dalian city in north-east China's Liaoning province (*Australian Journal of Mining* 1991). There are at least six diamond-producing districts in China, four being of the alluvial type and two containing primary sources hosted in Kimberlite. The alluvial deposits are located in the Tacheng and Linshu areas of Shandong province, Yuanshui River in Hunan, and Jingshan in Hubei province (Bureau of Mineral Resources 1988). Alluvial deposits are also known to occur in extreme Xinjiang and in Anhui province. The two Kimberlite-hosted diamond districts in China are located in Liaoning province (Fuxian county) and Shandong province (Mengyin county).

Since the Wafangdian mine commenced production in 1990, several diamonds weighing more than 10 carats have been found. The Liaoning deposit (Wafangdian site) makes up 52 per cent of China's known reserves. The mine has reportedly been negotiating with the United States, Belgium, and Hong Kong companies to establish a joint venture for diamond processing. China currently imports 90 per cent of its diamond requirements.

China has identified reserves of six mineral fuel resources, including coal, crude oil (petroleum), natural gas, oil shale, uranium, and thorium. The nation is one of the wealthiest coal-bearing countries in the world and may expand its resources of crude oil and natural gas through both onshore and offshore exploration. Among the six fuel resources, the output values of coal and petroleum (refined products) are by far the most significant.

Coal is abundant in China. Total reserves (anthracite, bituminous, and lignite) amount to approximately 114.5bn. tonnes, which represent 11 per cent of the world's total (British Petroleum Company 1984–92 (June 1992): 26). Although coal occurs in almost every Chinese province and autonomous region, over 85 per cent of the coal reserves are located to the north of the Qinling Mountains (Shaanxi province) and the Huai River (Anhui and Jiangsu provinces), and about 60 per cent in north China. Shanxi and Inner Mongolia boast the largest coal reserves in China; collectively these two provinces contain approximately one-half of China's proven reserve totals. North-western, south-western, and north-eastern China each account for about 10 per cent of the national total (Hu, Zhang, and Liu 1984: 72).

China's coal resources represent a complete range of types and qualities, including anthracite and various kinds of bituminous coal, coking coal, and lignite. Two-thirds of China's coal reserves are bituminous coal, of which a relatively small proportion is coking coal used by China's metallurgical industry for steel-making. The prominent reserves of hard coal are of bituminous ranks, and most of these reserves have a sulphur content of less than 1 per cent. In general, coal reserves to the north of the Yangtze River are of high quality.

In addition to coal resources, China is well endowed in oil and gas. According to the US Bureau of Mines, China has 5.4m. square kilometres of oil- and gas-bearing sedimentary rock basins, which represents more than half of the nation's vast land area. At the end of 1991, China's proven oil reserves were estimated at 3.2bn. tonnes, which represented 2.4 per cent of the world's total (British Petroleum Company 1984–92 (June 1992): 2). Natural gas proven reserves were estimated at 920.4m. tonnes oil equivalent, or 0.8 per cent of the world amount (British Petroleum Company 1984–92 (June 1992): 18). The Ministry of Geology and Mineral Resources states that as much as 1m. square kilometres of land in China has not yet been geologically surveyed in detail for hydrocarbon-bearing sequences (Gao 1986: 4) and, as a result, significant oil and gas resources may still be discovered.

As for offshore resources, because China has nearly 1.4m. square kilometres of continental shelf at a water depth of less than 200 metres, prospects for offshore oil discoveries are somewhat promising despite disappointing results in early commercial exploration activities. China possesses a very large coastline, with Bohai in the north, the Yellow Sea and the East China Sea in the east, and the South China Sea in the south. During the early 1980s, China's offshore oil regions were considered among the most promising throughout the world. However, by 1986 many oil companies had come to realize that the South China Sea—on which exploration activities were focused—was unlikely to contain large oil reserves because of less-than-favourable geologic conditions and low gas content of the oil (*South China Morning Post* (1986)). During the years following the first and second rounds of bidding for offshore oil (held in 1982 and 1985 respectively), total investments amounted to $2.42bn. and 400m. yuan. Despite such efforts, oil discoveries to date have been only modest, and offshore production may rise to around 8m. tonnes per year by the late 1990s.

GEOGRAPHIC DISTRIBUTION OF MINERAL SUPPLIES

Table 4.2 lists China's major mineral and energy deposits and mines, smelting and refining centres, and semi-manufacturing/manufacturing centres by province. This information has been compiled from several recent Chinese and Western publications, including Ministry of Geology and Mineral Resources (1991*a*, *b*), Energy, Mines, and Resources Canada (1988), US–China Business Council (1988), China National Nonferrous Metals Industry Corporation (1986), and Ikonnikov (1975). Only the principal deposits and mines are listed as there are virtually thousands of additional small-scale deposits and mines in the nation that are not accounted for in the table.

As a means of visually depicting the information presented in Table 4.2, Figures 4.1, 4.2, and 4.3 are included here. These figures illustrate the major mineral and energy deposits of China, along with the nation's processing and refining centres. As shown, the regional distribution of China's mineral and energy supplies is uneven and does not necessarily correlate with the localities of existing consuming centres. This imbalance places severe strain on China's transportation network and influences the distribution of heavy industry.

The majority of non-fuel mineral deposits in China are concentrated in the east, although this is in part a function of exploration biases which were prevalent in the past. Non-ferrous metal reserves are, in general, located near energy sources such as hydropower, petroleum, or coal, making energy-intensive smelting activities convenient. As an example, two of China's largest metal deposits—Lanping (lead/zinc) and Pingguo (bauxite/alumina)—are located in Yunnan province and Guangxi autonomous region respectively, where rainfall and numerous rivers provide for abundant hydroelectric resources. However, as with many mineral deposits in China, their localities are in remote regions with difficult mining conditions and poor transport, leading to rather slow exploitation.

According to a World Bank (1985*b*) study, the potential advantages of abundant energy close to mineral deposits have not been realized in China (with a few exceptions). The majority of processing facilities (smelters/refineries) in the nation are located in coastal provinces some distance from mineral reserves and energy sources. Nearly one-third of China's refined copper output, for instance, emanates from processing facilities in Shanghai and Liaoning, yet major copper deposits and mines are found in at least eight provinces/regions throughout the country.

Like non-fuel minerals, China's energy supplies are unevenly scattered throughout the nation. Although China has 18 major sedimentary basins containing oil and gas, 85 per cent of proven crude oil reserves in China are in the eastern part of China and north of the Changjiang River (Xin Dingguo 1988). Similarly, proven coal reserves are also unevenly distributed, with an estimated 78 per cent located in north and north-west China (Xin Dingguo 1988). Of the total coal reserves, Shanxi province and Inner Mongolia contain 56.4 per cent, the south-west 9.6 per cent, and east China holds only 6 per cent.

The principal energy-consuming regions of China are in the east, north-east, and south-central parts of China. These areas are, in general, heavily populated and economically developed. They are also typically located some distance away

TABLE 4.2. *Distribution of mineral and energy supplies by Chinese province/region*[a,b]

Province/region	Principal deposits and mines	Principal smelting/refining centres	Principal semi-manufacturing/manufacturing centres
Anhui	Coal, copper, fluorspar, iron ore, phosphate rock and apatite, salt, sulphur, vanadium	Aluminium, copper	Cement, clays, copper, salt, steel
Beijing	Coal, iron ore, kyanite and related materials, silver	Platinum	Cement, petroleum, steel
Fujian	Bauxite, fluorspar, iron ore	Lead, uranium,[c] zinc	Cement, clays, titanium
Gansu	Cobalt, copper, fluorspar, iron ore, lead, magnesite, nickel, petroleum, platinum, silver, zinc	Aluminium, asbestos, cobalt, copper, lead, nickel, platinum, REE, zinc	Cement, copper, steel
Guangdong	Fluorspar, gypsum, iron ore, lead, manganese, mercury, phosphate rock and apatite, silver, sulphur, tungsten, zinc	Lead, REE, tin, tungsten, zinc	Cement, petroleum, steel, titanium
Guangxi	Antimony, barite, bauxite, cobalt, fluorspar, gold, cobalt, fluorspar, gold, gypsum, manganese, REE, silver, sulphur, talc and related materials, tin, titanium	Alumina, aluminium, tantalum (by-product of tin-smelting), lead, tin, tungsten, uranium,[c] zinc	Cement, talc and related materials
Guizhou	Barite, bauxite, coal, gypsum, mercury, molybdenum, phosphate rock and apatite	Alumina, aluminium, mercury, titanium	—
Hainan	Cobalt, graphite, iron ore, natural gas,[d] petroleum (offshore), REE, salt, titanium, uranium	Alumina	Steel

TABLE 4.2. (*cont.*)

Province/region	Principal deposits and mines	Principal smelting/refining centres	Principal semi-manufacturing/manufacturing centres
Hebei	Asbestos, bauxite, coal, fluorspar, gold, iron ore, kyanite and related materials, molybdenum, natural gas,[d] petroleum, phosphate rock and apatite, vanadium	Aluminium, asbestos, magnesite	Cement, steel
Henan	Bauxite, chromium, coal, fluorspar, gold, graphite, natural gas,[d] petroleum, silver, sulphur	Alumina, aluminium, copper, gold, silver	Cement
Heilongjiang	Coal, gold, graphite, molybdenum (Wudaoshan), natural gas,[d] petroleum	—	Cement, graphite, natural gas, petroleum
Hubei	Barite, coal, copper, fluorspar, graphite, gypsum, iron ore, natural gas, petroleum, phosphate rock and apatite	Aluminium, copper, lead	Cement, natural gas, steel
Hunan	Antimony, bauxite, chromium, coal, cobalt, fluorspar, gold, graphite, gypsum, lead, manganese, molybdenum, natural gas,[d] nickel, petroleum, phosphate rock and apatite, REE, silver, sulphur, tungsten, uranium, zinc	Aluminium, antimony, cadmium, cobalt, lead, molybdenum, tin, tungsten, uranium,[c] zinc	Cement, copper, natural gas, steel, uranium
Jiangsu	Coal, iron ore, kyanite and related materials, phosphate rock and apatite, salt	Asbestos	Cement, petroleum, sodium carbonate (natural and synthetic), steel, vanadium

Jiangxi	Barite, columbium and tantalum, copper, fluorspar, molybdenum, nickel, REE, salt, silver, tungsten, uranium	Copper, REE, tin, tungsten	Copper, tungsten
Jilin	Gold, graphite, nickel, petroleum, platinum	Aluminium	Petroleum
Liaoning	Asbestos, bauxite, bentonite, chromium, coal, copper, gold, iron ore, lead, magnesite, manganese, molybdenum, natural gas,[d] petroleum, salt, talc and related materials, zinc	Alumina, aluminium, asbestos, copper, gold, lead, platinum, titanium, uranium, zinc	Cement, copper, molybdenum, natural gas, petroleum, steel, talc and related materials, tungsten, vanadium
Nei Mongol	Chromium, coal, fluorspar, gold, graphite, gypsum, iron ore, petroleum, REE, salt, sulphur	Aluminium, REE	Cement, graphite, steel
Ningxia	Gypsum	Aluminium	—
Qinghai	Asbestos, lead, lithium, potash, salt, zinc	Aluminium, magnesium metal	Salt, titanium
Shaanxi	Antimony, asbestos, coal, gold, gypsum, mercury, molybdenum, platinum	Aluminium, asbestos, platinum	Cement, steel
Shandong	Barite, coal, fluorspar, gold, graphite, iron ore, magnesite, natural gas,[d] petroleum, phosphate rock and apatite, REE, salt (Huang Hai), silver, talc and related materials	Alumina, aluminium (natural and synthetic)	Cement, sodium carbonate (natural and synthetic), natural gas, petroleum, talc and related materials, graphite, steel
Shanghai	Iron ore	Asbestos, cobalt, lead, copper, REE, titanium, zinc	Cement, copper, petroleum, steel (Baoshan), vanadium
Shanxi	Bauxite, coal, copper, gold, graphite, gypsum, iron ore, kyanite and related materials, molybdenum	Alumina, aluminium, copper	Cement, steel
Sichuan	Asbestos, chromium, coal, gypsum, iron ore, molybdenum, natural gas,[c] nickel, petroleum, phosphate rock and apatite, salt, silver, titanium, vanadium, zinc	Asbestos, uranium,[c] tungsten, vanadium	Cement, natural gas, steel, titanium, tungsten, vanadium

TABLE 4.2. (*cont.*)

Province/region	Principal deposits and mines	Principal smelting/ refining centres	Principal semi-manufacturing/ manufacturing centres
Tianjin	Natural gas, petroleum	Copper	Cement, copper, petroleum, steel, sodium carbonate (natural and synthetic)
Xinjiang	Chromium, cobalt (contained in Ertix He nickel deposit), columbium and tantalum, gold, kyanite and related materials, lithium, nickel, petroleum, platinum, salt, uranium	Uranium[c]	——
Xizang	Chromium, copper, lithium, salt	——	——
Yunnan	Cobalt, copper, gold, lead, phosphate rock and apatite, silver, tin, zinc	Aluminium, copper, lead, tin, zinc	Cement, copper
Zhejiang	Bentonite, fluorspar, gold, lead, silver, zinc	Uranium,[c] tungsten	Cement, clays, petroleum, steel

Note: —— No major deposit, mine, or centre listed. REE rare earth elements.

[a] Major deposits, mines, processing, and production facilities only.

[b] As of 1992.

[c] Possible.

[d] Associated with petroleum occurrence.

[e] Non-associated.

Sources: Minerals Policy Program (1993); Geological Publishing House (1992, 1990); Energy, Mines and Resources of Canada (1988); US–China Business Council (1988); Chin (1987*a*, *b*, 1986, and 1985); China National Nonferrous Metals Industry Corporation (1986); DeYoung, Sutphin, and Cannon (1984); DeYoung, Sutphin, Werner, and Foose (1985); Zhu G. (1983); and Ikonnikov (1975).

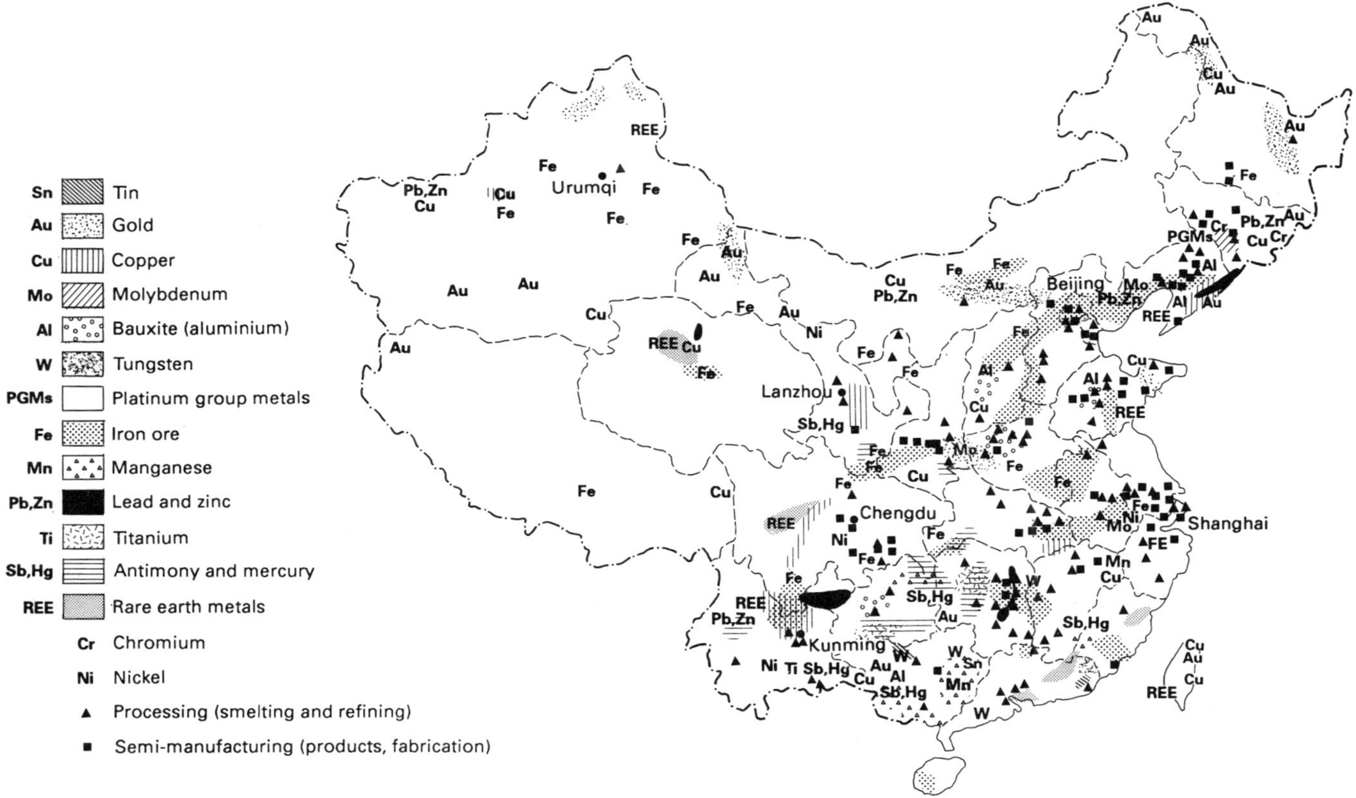

FIG. 4.1. China's mineral deposit areas and processing facilities

FIG. 4.2. China's oil and gas deposits, refineries, and petrochemical complexes

FIG. 1. China's coalfields and major production sites

from the areas where energy reserves are concentrated, causing problems in transportation.

The imbalance between fuel supplies and demand (consumption) in China is serious and has hindered economic and industrial development (described further in Chapter 7). Transportation bottlenecks are most severe in the energy sector, where limited rail capacity sometimes makes it impossible to move coal from mines to industrial and urban centres. Rail is the most important means of transporting coal from west to east in China, where more than 60 per cent of the nation's coal output is moved by rail (Jia Yunzhen, 1988*a*, *b*). To enhance rail transport of coal from north to south in China, the Qinghuangdao (Hebei) and Shijiusuo (Shandong) coal ports have recently been built or expanded. Qinghuangdao port, on the western shore of the Bohai Sea, is now undergoing modernization, and is expected to handle 120m. tonnes of coal and other cargoes by the year 2000. While some coal-mining districts in China are devoted to export, in other areas, particularly in the south, it is considered less expensive to import coal rather than to transport it across the mainland.

The energy problem of China has particularly plagued rural areas, where a huge portion of the total households are in need of additional fuel supplies. Rural energy has long been one of the most critical issues in China's national economic and energy development. Nearly 600 million people live in the rural regions of China, which cover millions of square kilometres in land area. Although China's rural economy has developed very rapidly in the past few years, the majority of the energy supplied to rural people is consumed in daily living activities. Commercial energy supplied to the rural regions amounts to about one-sixth of the national total, which satisfies up to 20 per cent of the total demand in rural areas (Xin Dingguo 1988). The remaining 80 per cent of rural energy depends exclusively on non-commercial resources, among which 90 per cent is biomass energy, including straws, stalks, and fuel woods.

Overcoming the difficulties and constraints caused by the separation of many mineral and energy deposits and mines from processing facilities will be a major challenge facing China's leaders. Substantial capital investment is expected to be directed towards the retrofitting and/or development of facilities near urban centres, creation of new rail lines and coal slurry pipelines, and high-voltage transmission lines. At present the Chinese government is encouraging the development of energy resources near consuming centres in order to reduce transportation bottlenecks.

MINING-RELATED INDUSTRIES OF CHINA

Production of metals, non-metals, and mineral fuels in China is controlled largely by government policies and objectives. Indeed, the Chinese government has a major supervisory role over the many mineral-related industries and corporations in the nation.[6]

The State Council of China is the executive body of the highest organ of state power in addition to being the highest organ of state administration. It is responsible to the National People's Congress or to its own permanent organ—the Standing Committee—when the Congress is not in session. Major functions and powers exercised by the State Council include leadership and guidance over ministries, commissions, and local state bodies, drafting and implementing national economic plans and the state budget, and directing and governing all state administrative affairs.

As part of an overall reorganization of the Chinese government in 1988,[7] the State Council was formally restructured. The plan for restructuring was adopted during the spring first session of the Seventh National People's Congress and involved a cut in the number of State Council ministries and commissions from 45 to 41. Although the number of ministries was reduced by only four, the reform involved was much more complex.

According to the plan, 12 ministries and commissions were abolished or merged, while nine new ones were established. The reform involved not just abolition, reduction, and amalgamation of the ministries and commissions, but a general change in the functions of many government organs. China's government reorganization was aimed at establishing a more flexible, efficient, and smoothly operating system of administration and management, and increasing the separation of government from economic management. Another major streamlining and reorganization of central government departments under China's State Council may take place in 1993/1994, as part of an ongoing drive to reduce bureaucracy and as a cost-cutting measure.

With restructuring in 1988, 41 ministries and commissions now exist under the direction of the State Council, including the Ministries of Metallurgical Industry, of Energy, of Materials and Equipment, and of Geology and Mineral Resources. Each ministry and commission has one minister and two to four vice-ministers. These ministers have the overall responsibility for their organizations.

Below the State Council is a vitally important organization—the State Planning Commission—which is responsible for long-term (five-year) planning, development of new capital projects, budgeting capital investments, and project financing and approval. In general, any large project involving more than $US10m. investment must be approved and financed by the State Planning Commission.

The State Economic Commission, also once below the State Council, was abolished in late 1988 as part of the attempt at market-oriented reforms and to reduce the scope of direct annual planning—the competence of the State Economic Commission. In 1991, the State Council established the Production Office to reassert greater control over manufacturing/industrial state-owned enterprises and current production. In 1992, this office was expanded into the State

Council's Economic and Trade Office, or ETO, to include trade as well. ETO, considered a new economic super-ministry, is expected to be ultimately converted into a commission, modelled after Japan's Ministry of International Trade and Industry.

Up till early 1992, China's State Planning Commission remained more influential than ETO, but by mid-year ETO had become much more powerful, owing in part to the rising political influence of its chairman Vice-Premier Zhu Rongji. Zhu is favoured by Deng Xiaoping, and was elected a member of the Politburo of the Standing Committee of the Chinese Communist Party in October 1992. Clearly, some of the powers previously enjoyed by the State Planning Commission are presently being lost to ETO. Preliminary indications are that in the years ahead the State Planning Commission as well as the Ministry of Foreign Economic Relations and Trade may be abolished.

The mining industry of China is extraordinarily large and complex, as exploration, development, processing, refining, distribution, and trade of metals, non-metals, and mineral fuels involves a multitude of industries and organizations. Mines in China are classified as either state-run or non-state-run, where state-run mines are supervised by the various ministries, or local (provincial or municipal) governments, and non-state-run mines include private (individual or family), collective, and provincial, district, or county mines. More than 130,000 state and non-state mines exist in China, with nearly 7,500 state-owned mines above the county level (Ministry of Geology and Mineral Resources 1991*a, b*). The remaining mines are small ventures operated by collective-groups and individuals, employing an estimated 5 million surplus rural labourers. Approximately one-sixth of the output value of China's mineral production comes from collective and individual mines. In 1985, for example, collective and individual (non-state) mines produced 15 per cent of the country's iron ore, 20 per cent of gold, 25 per cent of coal, 30 per cent of non-ferrous metals, half of manganese and sulphur-iron ores, and 70 per cent of building materials, and a considerable amount of non-metallic mineral products (Summary of World Broadcasts 1986).

Figure 4.4 presents an organizational structure of China's mineral-related industries, and illustrates the enormous industrial base responsible for mining and other related activities in China. The figure depicts China's new mining organizational structure adopted as part of the overall government reorganization in 1988. Further restructuring is likely in the years ahead, as part of a continuing effort to streamline operations and increase efficiency. The nation's mining industry has, however, already undergone a number of organizational changes in recent years. Since the early 1980s, the trend has been to replace larger conglomerate organizations with smaller specialized ones, particularly in the production and trading of mineral and energy commodities. This has caused much confusion both within China and among the international community about the extent of responsibilities of the various organizations. At the same time, overlapping responsibilities among many key bureaucracies have caused anger and enhanced competition.

In total, there are 15 ministerial-level (*Bu Ji*) administrative organizations today which are, to varying degrees, directly involved with China's mining industry. These organizations serve to plan, guide, and implement all mineral-related activities in the nation. In terms of authoritative powers, state bureaus or national corporations (*Fu Bu Ji*), bureaus (*Ju Ji*), divisions (*Si Ji*), and sections and departments (*Chu Ji*) lie below the ministerial-level organs.

Several of the organizations shown in Figure 4.4 are 'primary' industries; that is, mineral or energy raw materials serve as inputs into the industries for eventual processing into finished products. These industries include the Ministries of Metallurgical Industry, of Energy, of Chemical Industry, and of Materials, as well as the China National Nonferrous Metals Industry Corporation (CNNC), China National Petrochemical Corporation (SINOPEC), and the China National Chemical Import and Export Corporation (SINOCHEM). Because of their importance to China's minerals industry, five of the aforementioned ministries and corporations are described below.[8] Combined, these five organizations directly employ about 10 million workers in China. Since there are several other organizations in the Chinese government hierarchy that are directly, or indirectly, involved in minerals-related work, the actual number of workers involved in some capacity with the minerals industry of China greatly exceeds 10 million.

1. Ministry of Metallurgical Industry

The Ministry of Metallurgical Industry is one of two primary organizations in China involved in the production of metals, the other being the recently formed CNNC. As of 1983, China's Ministry of Metallurgical Industry delegated its responsibilities relating to non-ferrous minerals development to CNNC. Thus, despite its name, the Ministry of Metallurgical Industry is responsible only for iron and steel (and the ingredients for making steel), as well as gold. The separation of China's ferrous and non-ferrous sectors was aimed at improving efficiency and quality control in the nation's iron and steel enterprises, which are considered critical to the overall modernization programme.

The Ministry of Metallurgical Industry primarily conducts exploration for and mines iron ore, and supervises the nation's iron and steel mills. As the Chinese government relies on foreign funds to help boost steel production capacity, the Ministry of Metallurgical Industry set up the China International Iron and Steel Investment Corporation (CIISIC) in 1987 and empowered it to negotiate with foreign bankers, machine-builders, and iron and steel producers. According to the ministry, the Chinese steel industry since 1978 has utilized US $4.8bn. in foreign exchange for the introduction of advanced technology and facilities, in some 560 cases (Odagawa 1988).

At present, the ministry's main responsibility is to oversee China's steel industry, which consists of 14 large iron and steel complexes, about 1,800 plants (of which 1,246 are actual iron and steel mills) and 3.2 million employees (of which 2.58

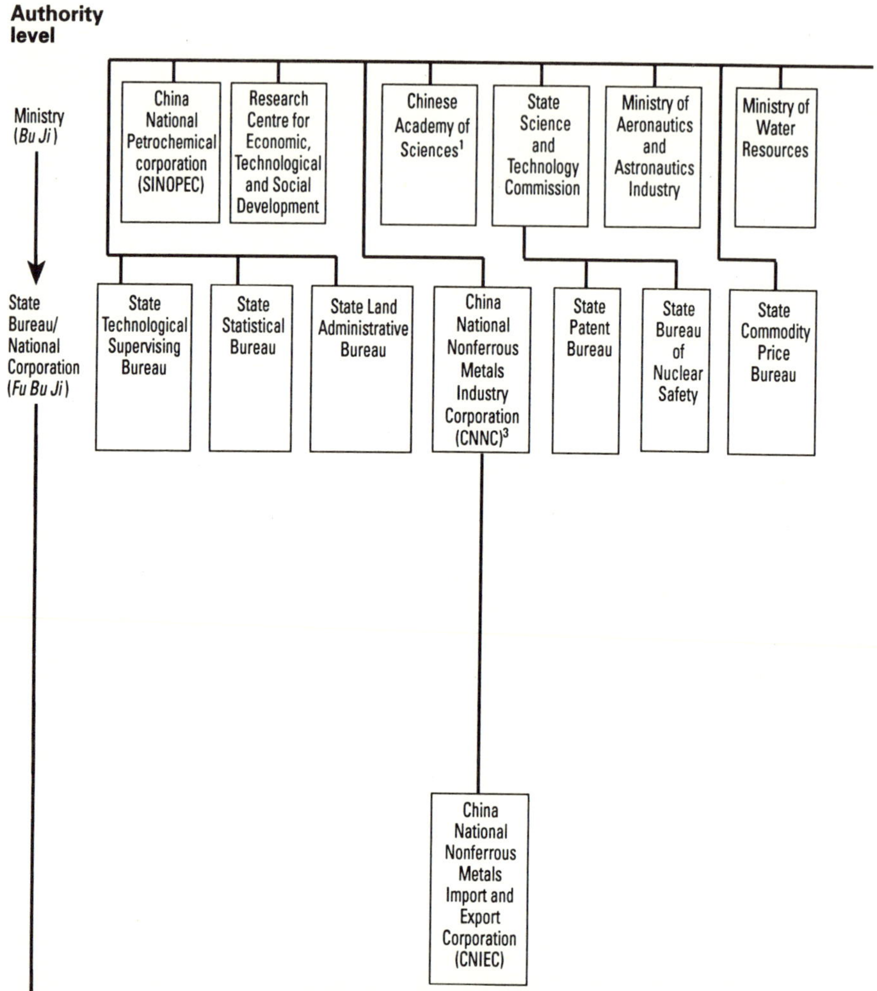

1 Sometimes referred to as Academia Sinica; composed of more than 80 research institutes

2 The committee is composed of senior officials from related departments under the State Council, including the State Planning Commission and Ministries of Metallurgy, Energy, Chemicals, and Water Resources. This committee was formed in 1953, but its work was suspended during the ten years of turmoil between 1966 and 1976. As of May 1984, the committee was reinstated and work resumed.

3 Level of authority not yet finalized.

4 Status now uncertain.

5 In keeping with the policy of separating adminstration from business, the Gold Bureau formed a separate, related company called the China National Gold Corporation to handle day-to-day commercial affairs. Many of the employees within the Gold Bureau are also employed within the China National Gold Corporation.

6 Established in 1987.

7 SINOCHEM exports and imports chemicals, petrochemicals, and refined petroleum products under the supervision of MOFERT.

Note: Level of authority below Bureau includes Division (*Si Ji*) followed by Section or Department (*Chu Ji*).

FIG. 4.4. Organizational structure of China's mining industry

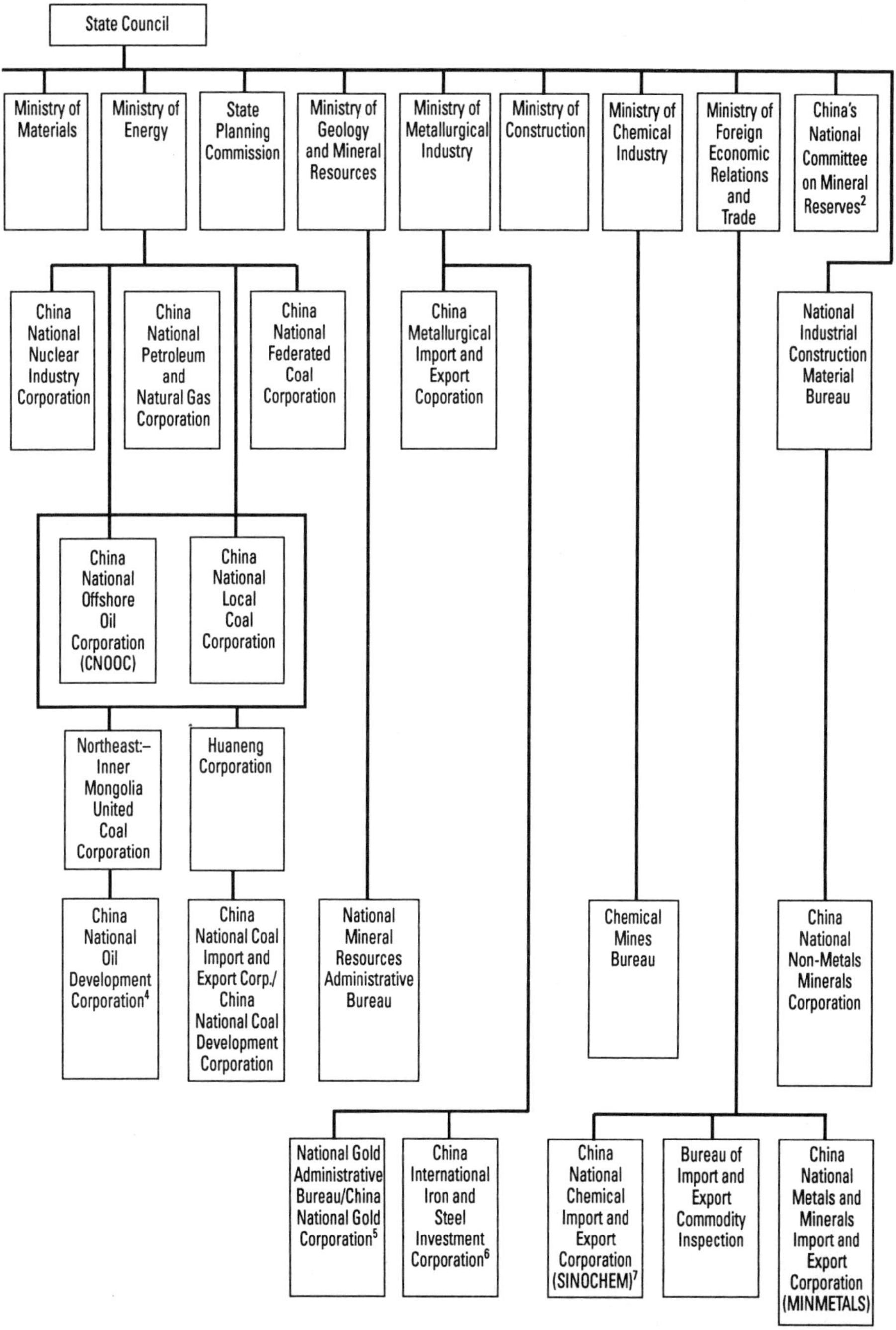
State Council
Ministry of Materials
Ministry of Energy
State Planning Commission
Ministry of Geology and Mineral Resources
Ministry of Metallurgical Industry
Ministry of Construction
Ministry of Chemical Industry
Ministry of Foreign Economic Relations and Trade
China's National Committee on Mineral Reserves[2]
China National Nuclear Industry Corporation
China National Petroleum and Natural Gas Corporation
China National Federated Coal Corporation
China Metallurgical Import and Export Coporation
National Industrial Construction Material Bureau
China National Offshore Oil Corporation (CNOOC)
China National Local Coal Corporation
Northeast:– Inner Mongolia United Coal Corporation
Huaneng Corporation
China National Oil Development Corporation[4]
China National Coal Import and Export Corp./ China National Coal Development Corporation
National Mineral Resources Administrative Bureau
Chemical Mines Bureau
China National Non-Metals Minerals Corporation
National Gold Administrative Bureau/China National Gold Corporation[5]
China International Iron and Steel Investment Corporation[6]
China National Chemical Import and Export Corporation (SINOCHEM)[7]
Bureau of Import and Export Commodity Inspection
China National Metals and Minerals Import and Export Corporation (MINMETALS)

million are employees of these steel mills). For decades the leaders of the Chinese government have considered the nation's steel industry as the key link to industrial development. The Chinese steel industry has recorded very steady growth in both quantity and quality since the late-1970s when reform of the economy and open-door policies were begun. The industry has registered significant success in reducing deficits, improving quality, conserving energy, and increasing surpluses. China plans to reduce steel-making energy consumption from 1.63 tonnes of coal per tonne of steel to 1.55 tonnes by 1995 (Tse 1992).

In recent years, the ministry's main projects have involved the renovation and expansion of existing major plants in Anshan, Wuhan, Baotou, Beijing, Benxi, and Maanshan. A major undertaking during the 1980s was the construction of the major Baoshan integrated steel complex located just north of Shanghai. Baoshan is China's most advanced iron and steel complex with a designed capacity of nearly 7m. tonnes of steel per year. Analysts suggest that China's steel industry advanced about 20 years through construction of the technologically advanced Baoshan complex. Currently, the Baoshan Iron and Steel Complex must import iron ore from Australia, Brazil, and India because at about 35 per cent iron content Chinese ore is not concentrated enough for this plant. The successful operation of Baoshan has accelerated the development of China's metallurgical industry. China's State Council recently ratified the third stage of the Baoshan Iron and Steel Complex in Shanghai. When the expansion project is completed in 1999, Baoshan will have increased annual capacity from about 6.8m. tonnes currently to 10m. tonnes of molten steel (*Mining Journal*, 25 Sept. 1992). The proposed renovation includes a new blast-furnace, a continuous hot rolling mill, and three continuous cold rolling mills.

Besides exploration, mining, and processing activities, the Ministry of Metallurgical Industry is engaged in trade through its arm CMIEC, or the China Metallurgical Import–Export Corporation. CMIEC was formed in 1980 and, at that time, was the first company to compete with the only other metals-trading group—MINMETALS of the Ministry of Foreign Economic Relations and Trade (MOFERT). As the trading arm of the Ministry of Metallurgical Industry, CMIEC participates in the exportation and importation of iron ore, pig iron, steel and ferrous alloys, and refractory products, plus the rare earths products produced at Bayan Obo and other Ministry of Metallurgical Industry facilities. The trade of many of these commodities was once handled almost exclusively by MINMETALS.

In addition to overseeing China's steel industry and associated ferrous mines, the ministry supervises gold-mining activities in China. Although state-sponsored exploration for gold is carried out by the Ministry of Geology and Mineral Resources, primary gold-mining activities are controlled by a central Gold Bureau under the Ministry of Metallurgical Industry. The Gold Bureau, an administrative entity within the ministry, formed a sister company to handle business affairs—the China National Gold Corporation. The main task of this corporation, which is also under the jurisdiction of the Ministry of Metallurgical Industry, is to work with foreign companies. Major gold-mines at the provincial level periodically report to the Gold Bureau/China National Gold Corporation.

Some of the administrative duties and responsibilities related to gold development in China are shared by the recently established State Gold Administration of China. Formed in 1989, this administrative body reports to the Ministry of Metallurgical Industry, and is responsible for the central administration of the nation's gold industry. The State Gold Administration will help the Ministry of Metallurgical Industry facilitate expanded output of gold in China, which will be used to help the country in the repayment of its foreign debt. China's foreign debt is expected to peak in the early 1990s, at which time gold production is scheduled to be nearly twice the level of the mid-1980s.

2. *China National Nonferrous Metals Industry Corporation (CNNC)*

In April 1983, the China National Nonferrous Metals Industry Corporation (CNNC) was established under the State Council to oversee China's non-ferrous metals industry. The formation of CNNC was deemed necessary as a result of the shortage of non-ferrous metals (particularly copper and aluminium) in China and their increasing importance to the nation's industrialization, and a growing non-ferrous metals import bill. Since 1983, China's total annual output of ten major non-ferrous metals—copper, aluminium, lead, zinc, nickel, tin, antimony, mercury, magnesium, and titanium—has ranked among the top six countries in the world. China is today the world's fourth largest non-ferrous metal producer. There has been significant growth in the variety of non-ferrous metals produced, as well as in rare earth metals, precious metals, and semiconductor materials.

CNNC operates 15 regional companies which control production units directly or in conjunction with provincial authorities. The corporation produces two-thirds of China's annual total of non-ferrous metals. CNNC has approximately 1.2 million employees, of which about 100,000 are professionals. It supervises the activities of more than 900 non-ferrous metals enterprises and institutions throughout China, and with the exception of Tibet (Xizang), all provinces and autonomous regions in the country have constructed their own local non-ferrous metals industries (Huang Jichun 1987).

Responsibilities of CNNC are multifold and include geological studies and surveys, mineral assessments, mining development, smelting and refining, manufacturing, research and education, and metals-trading. It also evaluates government principles, policies, and regulations underlying development of the non-ferrous industry, making sure they are implemented after approval by the State Council. As for trading, in early 1984 the China National Nonferrous Metals Import and Export Corporation (CNIEC) was formed to serve as the trading arm of CNNC. CNIEC's principal responsibilities include the exportation of mineral ores, finished products, semi-metal equipment, and technology, as well as the importation of ores, metals and finished products, machinery and complete plants, instruments, spare parts, and technology. The range of minerals traded through CNIEC includes the non-ferrous metals and rare earths. In recent years CNIEC and MINMETALS have competed strongly against one another, although CNIEC has steadily grown in importance, now controlling many of China's large non-ferrous mines and processing facilities. There remains stiff competition between CNIEC and MINMETALS at the provincial and local levels, where few regulations dictate which corporations handle specific metals and metal products.

Although the Ministry of Geology and Mineral Resources is China's principal organization involved in exploration activities, CNNC also conducts exploration for the same minerals at times, although their general focus is different. The ministry often explores in new or remote regions, whereas CNNC generally undertakes exploration in previously explored regions where known deposits exist.

3. *Ministry of Energy*

As part of China's government restructuring in mid-1988, a new Ministry of Energy was established. Composed of the former ministries of petroleum, coal, and nuclear industries, and part of the former Ministry of Water Resources and Electric Power, the newly established Ministry of Energy has the overall responsibility of co-ordinating the nation's energy development and utilization—a difficult task in view of China's huge energy requirements and need of export earnings. Several million workers are employed in the Ministry of Energy.

The ministry is composed of two major parts—administrative departments and operating corporations. The various administrative departments are concerned with individual energy forms as well as other functional areas such as investment, financing, technology transfers, and personnel management and training. The operating corporations have been established as independent economic entities, assuming full responsibility for their own profits and losses. The rationale behind this is to separate the policymaking role from the operational role. The corporations include the China National Offshore Oil Company (CNOOC), the China National Nuclear Industry Corporation, the China National Federated Coal Corporation, the China National Petroleum and Natural Gas Corporation, the Northeast–Inner Mongolia United Coal Corporation, the China National Local Coal Corporation, and the Huaneng Corporation. These seven state corporations have equivalent status and are independent. The China National Petroleum and Natural Gas Corporation undertakes the responsibilities of the former Ministry of Petroleum Industry, including the exploration, development, and production of onshore oil. This corporation and CNOOC are responsible for their own financial profits and losses. The China National Local Coal Corporation supervises the nation's private and individual coal-mines—those that were under the authority of the former Ministry of Coal Industry.

In addition to the aforementioned operating corporations, the China Electricity Association is part of the Ministry of Energy, serving as a research co-ordinating organ. The responsibilities for water resources and electricity development in China were separated with the division of the former Ministry of Water Resources and Electric Power; electricity development issues are now under the mandate of the Ministry of Energy. China's electricity system is divided into several regional grids—with each grid essentially operating as an electricity utility.

4. China National Petrochemical Corporation (SINOPEC)

In 1982 China's State Council established a working group to negotiate with the major players involved in petroleum-refining in China—namely the industries of petroleum, textiles, chemicals, and various local governments. At stake was the pending formation of a new ministerial-level economic entity to be responsible for crude oil refining, marketing, and export—highly profitable industrial activities. The negotiations resulted in the establishment of the China National Petrochemical Corporation, known as SINOPEC.

Formally established in July 1983, SINOPEC was made a licensed profit-making corporation directly beneath the State Council. The Ministry of Petroleum and other players involved in the negotiations which led to the formation of SINOPEC were reluctant to turn over many profitable responsibilities to SINOPEC; hence SINOPEC's mandate was reduced even before formation. For instance, several small oil field refineries (less than 20,000 b/d in capacity) were retained by the Ministry of Petroleum, and 10 per cent of refinery profits were earmarked for local governments (Fridley and Fesharaki 1987*a*). Perhaps the most disputed compromise was allowing product exports to remain in the control of SINOCHEM, the crude and petroleum product trading arm of MOFERT (Fridley and Fesharaki 1987*a*). With the enormous foreign exchange earnings from product exports at stake, the ministry fought bitterly to maintain its 30-year monopoly on exports.

Today, SINOPEC manages nearly 95 per cent of China's petroleum-refining industry and is responsible for domestic product and petrochemical distribution. The organization consists of 72 units including 38 giant and large oil-refining, petrochemical, synthetic fibre, and chemical fertilizer enterprises (under which there are 37 refineries, 21 basic organic chemical plants, 15 synthetic fibre plants, 5 synthetic rubber plants, 3 synthetic resin plants, 8 plastics plants, and 13 chemical fertilizer plants), as well as units engaging in foreign trade, marketing, research and development, design, and construction as well as colleges and institutions. Moreover, the Corporation has set up branches and representative offices in Japan, the United States, Germany, Thailand, Ecuador, and Hong Kong. More than 50 joint ventures of different scale have been established. In collaboration with Japanese firms it has set up the Shihua International Leasing Co. Ltd. and other joint ventures. By the end of 1990, the staff and workers totalled 630,000, including 75,000 technical persons. In addition, there were 200,000 staff and workers in the sales sector.

SINOPEC has 30 refineries with crude distillation capacity above 2.5m. tonnes each, and total crude distillation capacity of 139m. tonnes accounting for 89.8 per cent of the country's output (China Petrochemical Corporation 1991). The secondary processing capacity is over 62m. tonnes. There are five ethylene production bases, each having an annual production capacity of over 300,000 tonnes, and a total capacity of 1.802m. tonnes, accounting for 87.1 per cent of the country's total. The Corporation annually provides markets with more than 1,500 oil and chemical products totalling approximately 90m. tonnes. The present storage and transportation decks belonging to enterprises situated along sea coast and riverside have 77 berths with an annual handling capacity of over 90m. tonnes. Except for Tibet and some other regions, every county of the country has its own service stations, the total number exceeding 5,000.

5. *Ministry of Geology and Mineral Resources*

Like the other mineral-related ministries in China, the Ministry of Geology and Mineral Resources reports directly to the State Council. It is China's principal exploration-oriented state organization, searching for all mineral and energy resources throughout the country. The ministry employs approximately 420,000 workers, of which nearly one-quarter are geoscientists. In terms of organizational breakdown, the ministry is composed of several bureaus and departments, as well as research institutions and universities. The major Wuhan College of Geology (formerly the Beijing College of Geology) is under the supervision of the Ministry of Geology and Mineral Resources. It has over 6,000 students alone and consists of seven major geoscience and engineering departments.

Until 1984, the Ministry of Geology and Mineral Resources had no production facilities or operating mines; these functions were relegated to China's production-oriented organizations, including CNNC and the Ministry of Metallurgical Industry. However, with a relaxation of government regulations and a general decentralization mindset, in 1984 the ministry was given the right to own and operate some of its own mines and production facilities (US–China Business Council 1988). Since then the ministry has operated a few production facilities, particularly those involving non-metallic mineral resources. With these operations the ministry is able to reap some of the revenues resulting from its own exploration work—an added financial incentive to improve exploration strategies and techniques. Efforts are continuing within the ministry to broaden its involvement in development activities, as a special department is now being formed which will be responsible for seeking joint-venture mining projects in China. This department will be essentially self-ruling and act much like a mining corporation. The changing forms of economic co-operation and exchanges the Ministry of Geology and

Mineral Resources has pursued with foreign partners over the past four decades have been closely linked to the changing policies of the government. Since 1949, the ministry has engaged in scientific exchange with 79 countries, signed 38 agreements with 20 countries, joined 28 academic institutions, and maintained contact with 12 organizations under the auspices of the United Nations (*China Daily*, 27 Nov. 1989).

The fundamental responsibility of the Ministry of Geology and Mineral Resources is to survey and explore China's surface terrain, onshore as well as offshore, for natural resource occurrences. Other ministerial functions include basic geological, geochemical, and geophysical research, and short-term to long-term mineral resource planning. Exploration is commonly broken down into four stages—mineral resource estimation, prospecting (of geology, geophysics, and geochemistry), geological exploration (drilling), and identification and characterization of mineral and energy reserves. Surveying is intended to facilitate the development of detailed geological, geophysical, and geochemical maps needed to assist geologists in the discovery of ore deposits. Map coverage of China at a 1:200,000 scale is a long-term objective of the ministry.

The Ministry of Geology and Mineral Resources unveiled in November 1989 a 12-year surveying programme aimed to extract enough resources to support the country's economic growth. For energy-related mineral resources, according to the plan, the priority will be on prospecting for oil and natural gas while continuing to search for additional coal and large high-grade uranium reserves. Proven oil reserves identified in these 12 years, it is hoped, will equal the quantity discovered in the past forty. The target for natural gas is 5.8 times as much as in the last four decades.

Combined, the five mineral-related ministries and corporations described in the foregoing paragraphs directly employ at least 10 million workers in China. These ministries and corporations are directly and exclusively involved in minerals-related research, production, sales, or trade activities. As was depicted in Figure 4.4, there are several other organizations in the Chinese government hierarchy that are directly or indirectly involved with minerals-related work. As a result, significantly more than 10 million workers are involved in varying extents with the mining industry of China.

With decentralization during the 1980s and persistent efforts by China's government to distance itself from enterprises, the organizational structure of China's mining industry has changed significantly. Reorganization meant increased specialization, changing duties and responsibilities, overlapping interests, and often confusion. The overlapping interests among so many organizations enhanced a competitive spirit which was hoped by the Chinese government to lead to greater efficiencies in operations. Though it is too early to judge the successes and failures of the decentralization of China's mining industry, problems did arise with the transfer of authority to provincial and local enterprises which had little experience in management or in dealing with foreign enterprises. Enhanced competition among provincial companies led to selling confrontations where individual provinces attempted to outsell other provinces by offering goods to the market at a lower price. Corruption also resulted from the decentralization of authority. With the continued evolution of economic reform in China, the nation's mining industry will also continue to change in structure, control, content, and focus.

DECENTRALIZATION AND ENTERPRISE CONTROL

The organizational structure presented in Figure 4.4 refers primarily to the state-owned or state-controlled enterprises engaged in China's mining activities. China's state-controlled ministries and corporations have operating enterprises at all levels, or specifically central ministerial, provincial, and local ranks. As for the country's mineral and energy industries, the distinction between central and provincial/local-level enterprises became more pronounced with decentralization—a fundamental policy promoted as part of economic reform. As local-level enterprises grew in size, responsibility, and wealth, they became increasingly influential in production and trade activities that previously would have been decided and planned only at the ministry level in Beijing. The distinction between central and provincial/local-level enterprises became vitally important to foreign companies hoping to invest in Chinese mineral projects. Because local-level enterprises usually have different motives, objectives, and skills from their parent enterprises, problems arose as a result of decentralization.

Under economic reform, non-state industries have grown at a much faster pace than state enterprises, while also becoming more profitable. Economic losses among state enterprises have indeed accelerated in recent years, and non-state industry now accounts for more than 48 per cent of gross industrial output in the nation. In 1978, just before China's open-door policies began, the state-owned firms accounted for 78 per cent of the nation's industrial output. The state share has dropped by more than 2 per cent a year ever since, and is now just over half (*The Economist* 1992). Chinese government officials estimate the figure will reach 25 per cent by 2000.

One of the outcomes of a more powerful provincial/local-level enterprise system in China is the elimination of complete and total authority by central ministerial-level enterprises and corporations over their subsidiaries. In mining, economic reform has allowed non-state private and collective mining enterprises to flourish in China and gain in their relative share of national output. Private and collective mining activities now account for approximately 36 per cent of total ore production and an estimated one-sixth of production value (Ministry of Geology and Mineral Resources 1991*b*). Non-state enterprises have the flexibility of being semi-independent from the state in decision-making. Their proliferation has played a positive role in alleviating mineral supply shortages in

China while developing local economies and changing the industrial structure of rural areas.

During the 1980s, although the state-controlled trading organizations MINMETALS and CNIEC did not compete very intensely at the central government levels (owing to greater control of facilities by CNIEC), at the provincial and local levels their subsidiary enterprises competed very strongly. Owing to decreased authority, China's central government was unable to intervene in many competitive situations and assign specific markets and products to the different enterprises. The rapid expansion of trading enterprises at the provincial and local levels was particularly troublesome to China's central government and entire mining industry, resulting in the introduction of new measures in 1990 to recentralize control over the trade system (further discussed in Chapter 5 below).

Trading companies at the local level complicated China's metals and non-metals trade by charging different prices for the same export products, where differentials in selling price for identical commodities were reported to be as high as 20 per cent (US–China Business Council 1988). The Chinese government requires the issue of an export licence before selected commodities (including fluorspar, magnesium, alumina, barite, rare earths, talcum, and graphite) can be sold overseas, yet some producers have been able to acquire export licences directly from provincial governments and export their own products abroad (sometimes at varying prices) without assistance from the ministry-level trading corporation (US–China Business Council 1988). Such practices have disrupted international mineral markets, such as that for tungsten in the mid-1980s. In this particular case, because of the large quantities of Chinese tungsten exports between 1983 and 1985, the world's market was flooded during these years, causing sharp declines in prices. As a means of preventing further disruption, the Chinese government subsequently asked MINMETALS and CNIEC to jointly form an organization (now referred to as the China National Tungsten Export Committee) to supervise all future tungsten exports from China.

The distinction between central government, ministerial-level enterprises, and their provincial/local subsidiaries extends beyond trade and production practices and affects foreign investment in China's mining projects, as well as efforts to raise capital. Decentralization has provided provincial/local mining enterprises more ability to solicit funds abroad for development projects, whereas in the past decisions to raise foreign capital were made primarily at the central government level. As economic reform has proceeded, China's government has reduced its allocation of funds to state-controlled enterprises and encouraged them to solicit loans from domestic banking institutions (World Bank 1990*a*, 1989). However, with the current fiscal clampdown, the government has frozen many new bank loans. The reduced number of bank loans still available are essentially for state-run enterprises, not private or collective operations.

As for foreign investment into China's mining projects, overseas companies participating in Chinese business ventures can presently deal with enterprises at the central ministerial, provincial, or local levels. This division is particularly important for equity participation, in the light of recent decisions by the Chinese government regarding the autonomy of ministry-level enterprises. Owing to the severe shortage of hard currency in China, these enterprises have been discouraged from engaging in joint-venture activities where a substantial amount of equity participation is involved. In the mining industry, for example, aside from projects involving iron and steel, only the Great Wall Aluminum Complex joint-venture project is expected to move ahead with significant ministry equity participation. This situation will remain unchanged in the near future as constraints on hard currency will likely grow as China is entering a high debt, repayment period.

While central ministerial-level enterprises have been discouraged from participating in equity joint ventures, provincial and local enterprises have not been so discouraged as of yet. Consequently, foreign companies interested in joint ventures in China can negotiate with these enterprises. It is advisable, however, that foreign companies notify first state enterprise officials in Beijing, then proceed with negotiations with relevant provincial or local enterprise officials.

From the perspective of a foreign investor, investment into provincial and/or local mining activities has several distinct advantages, including (1) central government interference can be minimized by working at the provincial and/or county level; (2) provincial and county governments often control the physical infrastructure for mines (including those of the state enterprises); (3) the provincial and/or local governments have tended to put forward smaller and better-quality ventures than the ministry-level enterprises—which traditionally propose marginal, highly capital-intensive ventures for the international investor; (4) many provincial and/or county mining ventures do not have to satisfy an output quota; and (5) provincial and/or local entities are generally less concerned with usage of state-of-the-art technology, but rather with an appropriate technology and more labour utilization, resulting in lower overall costs. In general, the time frame for implementing a project, the constraints on imports and exports of required equipment or product, and the level of bureaucratic interference can all be reduced when entering into ventures with provincial and/or county entities. This is particularly true in many autonomous regions, special economic zones, open coastal cities, and special planning provinces such as Guangdong and Hainan, which have special privileges in dealing with foreign investors.

There are, of course, some shortcomings and pitfalls in entering into joint-venture or other business activities with provincial or local enterprises that must be recognized and weighed against the advantages. These include the usual lower level of technical and/or managerial experience in initiating, developing, and carrying out a joint venture among the provincial and/or county entities. Therefore a certain amount of additional care must be taken in implementing a project to ensure that misunderstandings do not arise. On balance, however, a small-to-medium-size foreign investor in the mining industry of China must at least consider investment in provincial and/or county projects. Such investors must, however, pay attention to changing rules and regulations regarding authority and centralized control as they unfold in the 1990s.

RECENT MINERALS AND ENERGY OUTPUT OF CHINA

China is one of the world's wealthiest mineral-producing regions with a resource base highly capable of supporting a large industrial economy. The Chinese mining industry is the fourth largest in the world in terms of output value (Dorian 1987). It is a well-developed and robust industry, producing a wide variety of mineral and energy commodities. By 1992 Chinese ore mining output reached approximately 2bn. tonnes of various commodities, or more than 40 times that in the early 1950s. Although the range of commodities produced in China is extensive, there is a general lack of metal-processing facilities, necessitating large imports of aluminium, copper, magnesium, special steel products, titanium, and zinc, among other commodities.

China's recent minerals and energy production statistics are summarized in Table 4.3. The figures are estimates only, as the Chinese government does not release official production statistics for a majority of its mineral and energy commodities produced. Although all mineral and energy resources are the concern of this book, only production figures for selected metals, non-metals, and mineral fuels are shown in the table. The years covered in Table 4.3 include 1981, the beginning of the Sixth Five-Year Plan period, and 1989, 1990, and 1991. China's relative position of 1991 world minerals and energy output is presented in the final column of the table.

China continues to emphasize the development and expansion of its mining industry to facilitate economic growth. As a result, output of most mineral and energy commodities escalated during the decade 1981–91. Production of several mineral and energy commodities continued to rise in 1992. As for major metals, output of aluminium, copper, gold, iron and steel, tin, and tungsten rose substantially in the 1980s by at least a factor of one-third. China's largest metal-producing sectors are iron and steel, aluminium, and gold. Chinese crude steel output ranks fourth worldwide, behind the former Soviet Union, Japan, and the United States. As with aluminium, the current economic plan of China stresses the expansion and renovation of the domestic iron and steel industry.

Production of gold, iron ore, raw steel, tin, and tungsten in China accounts for a substantial share of world output, or 4.9, 18.6, 9.6, 12.5, and 5.2 per cent respectively, in 1991. Though China's crude steel output is large by world standards, per capita production (and consumption) of crude steel amounts to 45–50 kg, less than a third of the global average and only about 10 per cent of the level in developed nations (Odagawa 1988). The situation is similar for other metals like

TABLE 4.3. *Selected minerals and energy production of China, 1981–1991*[a] (tonnes '000, unless otherwise noted)

Commodity	1981	1989	1990	1991	% change: 1991 over 1981	1991 world rank	% of total
METAL							
Aluminium (metal, refined)	358.0	755.0	855.0	905.0	+152.8	7	4.7[b]
Copper (metal, refined)	300.0	470.0	490.0	560.0	+86.7	6	4.6
Gold[c]	52.9	100.0	100.0	110.0	+107.9	5	4.9[b]
Iron and steel:							
Iron ore[d]	110,000.0	145,000.0	168,317.0	170,000.0	+54.5	2	18.6
Steel, raw	35,600.0	61,200.0	66,038.0	70,570.0	+98.2	4	9.6
Tin (metal, refined)	16.0	28.3	28.0	28.0	+75.0	4	12.5
Tungsten (content)	13.5	21.0	21.0	21.0	+48.1	1	5.2
NON-METALS							
Cement (hydraulic)	82,900.0	210,290.0	209,710.0	248,000.0	+199.2	1	18.1
Fluorspar (acid and metallurgical grades)	480.2	1,700.0	1,500.0	1,500.0	+212.4	1	34.1
Gypsum					+158.5	3	8.2
Phosphate rock (P_2O_5 content)	3,443.0	8,900.0	8,800.0	8,900.0			
Salt	11,500.0	17,000.0	17,300.0	18,000.0	+56.5	4	11.3
	18,320.0	22,700.0	20,000.0	25,400.0	+38.6	2	13.7
MINERAL FUELS							
Crude oil	101,000.0	137,600.0	138,300.0	139,600.0	+38.2	5	4.5
Coal (excludes coke)	616,500.0	993,800.0	1,004,000.0	1,007,000.0	+63.3	1	22.9
Natural gas[e]	11.6	12.9	13.4	13.4	+15.5	22	0.7
Refined petroleum[f]	74,760.0	106,620.0	116,601.0	116,146.0	+55.4	5	NA

Notes: NA not available. Figures are rounded.

[a] Estimated.
[b] 1990 figure.
[c] Tonnes.
[d] In terms of 32.1% Fe ore, sometimes referred to as 'lean ore' in China.
[e] Million tonnes oil equivalent.
[f] Throughput.

Sources: British Petroleum Company (1984–92); US Bureau of Mines (1982–93); Mining Journal Limited (1982–92); World Bureau of Metal Statistics (1989–92); and State Statistical Bureau (1981–92).

copper and aluminium, as well as petrochemical products and many key industrial items.

China's tungsten output continues to dominate the world market, reaching an estimated 20,000 tonnes in 1991. However, its exports are primarily concentrates, not value-added products. The largest tungsten-producing province in China is Jiangxi, where the oldest operating mine at Xihuashan has an annual output of 2,500 tonnes of 65 per cent tungsten concentrate (US Bureau of Mines 1951–92 (1988)). Ten large tungsten mines are in operation in southern Jiangxi, which is considered the tungsten capital of China (US Bureau of Mines 1951–92 (1989)).

Gold production in China increased steadily to 1991, when the estimated output was 110 metric tons or 3,537,000 troy ounces. Chinese sources place the growth rate at around 11 per cent per annum over the decade 1981–91. Output of the precious metal was expected to increase steadily since 1978, when the decision to lift restrictions on collective and individual mining was implemented. At that time China changed its policy of allowing only state-owned enterprises to mine gold and instead adopted preferential policies to encourage people and production units to mine more gold. China has about 300,000 licensed private gold-miners today, but more people are mining the metal without permission. As a result of extensive smuggling of gold (estimated at 90 per cent of individually mined gold), wastage of ore, and unsafe mining practices by individual peasant miners in recent years, in November 1988 China banned individuals from mining gold, reversing the 1978 decision. The official price of gold was also raised at that time to the equivalent of $400 per ounce.

Like the metals, China's production of non-metals in recent years has grown rapidly, particularly that of cement, fluorspar, and gypsum. Output of each of these commodities nearly tripled since 1981. With China's ambitious industrialization programme receiving government priority throughout much of the 1980s, a dramatic boost in cement production was considered necessary. China's Sixth Five-Year Plan called for the construction of 25 large and medium-sized cement plants during the 1981–5 period, as well as the renovation of many existing plants. Remarkably, owing to substantial growth in rural and urban construction in China, cement production rose from 83m. tonnes in 1981, the first year of the Sixth Five-Year Plan, to 162m. tonnes in 1986, the initial year of the Seventh Five-Year Plan. Cement output continued to climb in the early 1990s, reaching 248m. tonnes in 1991, the first year of the Eighth Five-Year Plan. China's 1991 cement production accounted for 18.1 per cent of global output, ranking number one among cement-producing nations.

China's mineral fuels production is huge by world standards, as the nation's energy industry expanded significantly since the 1950s. Fuels are China's most important sector within the mining industry in terms of output value and quantity. Today, China represents the third largest producer of energy in the world.

Coal remains the backbone of China's energy economy, though development of its petroleum and natural gas resources is emphasized. In 1991, the nation's raw coal production reached 1.007bn. tonnes, which was 63.3 per cent greater than output in 1981. Output reached 1bn. tonnes in 1989 (including coke). China's present-day production of coal is the largest of any country worldwide, accounting for just under one-quarter of global production in 1991. Approximately 95 per cent of the production originates from underground mines, a figure that is higher than for any other major coal producer in the world.

Nearly 26 per cent of China's coal production comes from Shanxi province in north-central China. As with all other coal-producing provinces of China, Shanxi's output emanates from two basic categories of mines—state-operated and local mines. State mines are constructed using government investment and their production is distributed by the central government. Local mines can be further disaggregated into local state-owned, which are owned and administered by the provincial and county governments, and local collective or individually owned.

In 1990, aggregate coal production was contributed by some 618 state-administered mines (with an average output of 0.7m. tonnes), some 2,461 locally administered state mines (with an annual average output of 79,000 tonnes per mine) and an estimated 79,000 locally run small mines which contributed nearly 351m. tonnes (Ghose 1990). Much of the increase in coal production in the past decade has been contributed by the locally run small mines, whose share in the national output increased from 17 per cent in 1979 to 36 per cent in 1988. The state-run mines have also taken major strides in technological advancement, with the share of mechanized production increasing from 32 per cent in 1979 to a level of 58 per cent in 1988, amounting to 109.87m. tonnes.

As for the types of coal produced in China, bituminous coal makes up around 75 per cent of output, anthracite 20.7 per cent, and lignite 3.6 per cent. China has 14 large coal-mining areas with an annual output of more than 10m. tonnes, including Datong in Shanxi, Kailuan in Hebei, Pingdinshan in Henan, Yangquan in Shanxi, Hegang in Heilongjiang, Huaibei in Anhui, Xishan in Shanxi, Jixi in Heilongjiang, Xuzhou in Jiangsu, Fengfong in Hubei, and Fuxin in Liaoning. Aside from Shanxi province, Henan and Heilongjiang are major coal-producing provinces in China, accounting for 8 and 7 per cent respectively of national output. To ensure continued growth in the industry, China's Ministry of Energy is now attempting to shift coal production to the coal-rich north-west provinces of Shaanxi and western Inner Mongolia. China's largest production base sprawls across Shanxi, Shaanxi, and western Inner Mongolia. Under a development programme worked out by China's Ministry of Energy, this base should produce 545m. tonnes of coal in the year 2000, or 39 per cent of the country's output of 1.4bn. tonnes planned for that year (Du and Wang, July 1992). At that time, nearly three-quarters of the base's output, or 380m. tonnes, will be sent to other provinces.

Despite China being a major coal-producing nation, coal shortages occur throughout the huge country. The industry suffers massive losses, with stockpiles amounting to more than 300m. tonnes in 1991, 30 per cent higher than 1990 levels (*Mining Magazine*, Nov. 1991). To a large extent the current shortfalls are the result of inadequate infrastructure, and unreasonably low selling prices and high purchasing prices. The nationwide shortage has triggered 'speculation in coal' by industrial enterprises, which aggravates the already strained pricing and distribution systems. According to China's Ministry of Energy, 95 per cent of state-run coal producers, which account for nearly half of output capacity, are operating at a loss, and most large mines are mining at well below capacity. It is further estimated that 50 per cent of mines run by local governments are also loss-making.

A solution to China's menacing coal supply and distribution problems would involve better management, operations, and marketing co-ordination between the Ministries of Energy, of Supplies, of Railways, and of Communications, and a revamping of the pricing system. To this end, China's former Ministry of Coal Industry may be re-established to facilitate improved operations and a gradual conversion to market-oriented activities. The Ministry of Energy may also be partitioned or even abolished for the same reasons. In the interim, efforts will be made to accelerate price reform. The central government has already allowed the price of coal produced by several mines, or about half the country's output, to float at market levels. These sales were expected to reach 900m. tonnes in 1992 (*Mining Engineering* 1992). China's government is currently establishing coal wholesale markets in several locations.

China is very dependent on its coal output for consumption, and is becoming increasingly dependent on crude oil. As shown in Table 4.3, China's crude oil output reached 139.6m. tonnes in 1991, or 38.2 per cent greater than in 1981. Production of oil steadily increased during much of the past decade, in part owing to intensified development at China's major oil-producing districts—Daqing, Dagang, Shengli, Huabei, Liaohe, and Zhongyuan Basins. A majority of China's crude oil output is consumed domestically, with only marginal output being subject to export.

METALS AND ENERGY CONSUMPTION IN CHINA

Surging economic growth in China has relied largely on a high consumption of metals and energy resources, and the nation now accounts for a significant proportion of the world's total consumption. Excessive consumption of resources has resulted in excessive demand for even more resources and, in turn, the relative shrinking of China's resource base. Tables 4.4 to 4.7 provide data on China's recent consumption of metals and energy, including rates of growth or decline in consumption between 1981 and 1991. The tables present statistics for China, as well as its industrialized Asian neighbours, and the entire world, in order to provide some perspective on Chinese consumption levels. In examining any rates of growth of consumption, it is critical to consider also the sizes of the consumption bases being analysed.

TABLE 4.4. *Metals consumption in China and the world, 1991*[a] (tonnes '000)

Major metal	China	Japan	Korea, Rep. of	USA	World
Aluminium (primary)	800.0 (4.6)	2,431.6 (14.0)	383.3 (2.2)	4,200.7 (24.2)	17,342.7
Cadmium	500.0 (2.7)	4,990.0 (26.6)	400.0 (2.1)	3,238.0 (17.3)	18,732.0
Copper (refined)	590.0 (5.5)	1,613.2 (15.0)	343.3 (3.2)	2,124.5 (19.8)	10,754.6
Lead (refined)	250.0 (4.6)	422.2 (7.8)	173.0 (3.2)	1,246.8 (23.0)	5,413.9
Magnesium	15.5 (5.0)	27.2 (8.8)	1.7 (0.5)	91.9 (29.8)	308.7
Nickel	29.0 (3.4)	180.1 (22.3)	29.5 (3.7)	126.7 (15.7)	805.9
Steel, crude[b]	71,042.0 (9.7)	99,149.0 (13.5)	26,190.0 (3.6)	93,325.0 (12.7)	732,420.0
Tin	17.0 (7.7)	34.8 (15.7)	8.4 (3.8)	35.3 (15.9)	221.9
Zinc	530.0 (7.7)	845.5 (12.3)	283.5 (4.1)	933.2 (13.5)	6,888.0

Note: Figures are rounded.

[a] Figures in parentheses represent country's share of total world consumption.
[b] Apparent consumption (production plus imports minus exports).

Sources: Metallgesellschaft (1992); World Bureau of Metal Statistics (1989–92 (1992)); International Iron and Steel Institute (1993).

China's consumption of nine major metals in 1991 is large according to world standards (Table 4.4). In the case of crude steel, for example, consumption in China reached just more than 71m. tonnes in 1991, or 9.7 per cent of global consumption that year. Though not as large, this level of consumption is comparable to that of Japan, which reached 99.1m. tonnes in the same year. Japan, the most powerful economic force in Asia, boasts the second largest GDP worldwide behind the United States and one of the largest iron and steel industries. China's 1991 consumption of refined lead, magnesium, and zinc similarly approached quantities consumed in Japan.

Steel is the most significant metal consumed in China, accounting for more than 95 per cent of total metals consumption (by volume) in 1991. As a result of China's ambitious industrialization programme, consumption of steel is expected to rise significantly by the turn of the century, barring any major downturns in economic growth. In terms of domestic demand for finished steel, about one-third is used in heavy industrial applications, 12 per cent for transportation equipment, and 13 per cent for metal products (*Mining Journal*, 13 July 1990). During the 1980s, domestic demand for steel in China consistently outpaced consumption, though, while the nation is expected to be a net importer in the near future, assessing the likely level of imports is more difficult.

Recent studies completed in China and Australia (Chen *et al.* 1991; Shen 1988) found a strong correlation between steel consumption and National Income in the country.[9] Based on a vector autoregression model of the Chinese economy, assuming continued strong economic growth Chinese crude steel demand will grow from 60m. tonnes in the late 1980s to 100m. tonnes by the year 2000. This figure reflects expected decreases in steel usage due to substitution of other materials such as plastics and aluminium for steel, the influences of advanced technologies, a changing economic structure, and other factors. The Chinese steel market will then rank equal with markets in Japan, the United States, and the European Community. Steel demand is predicted to grow at about 5 per cent per annum in the second half of the 1990s. Fluctuations in steel consumption until the end of the century will probably be due to macroeconomic shocks that reflect political developments or fiscal and monetary responses that are policy-induced.

In addition to crude steel, China consumes relatively large amounts of aluminium, copper, lead, magnesium, tin, and zinc. In 1991 China ranked fourth in the world in consumption of tin, fifth for aluminium, copper, magnesium, and zinc, and eighth or ninth for lead, nickel, and cadmium. Consumption of aluminium was estimated at 800,000 tonnes, of which nearly half was imported. The Eighth Five-Year Plan of China emphasizes the expansion of aluminium production capacity in China to reduce the need for substantial imports of the commodity. Consumption of refined copper in 1991 reached 590,000 tonnes, with domestic metals production constituting most of the supply. China's consumption of aluminium, copper, cadmium, nickel, and zinc in 1991 continued to rise from the previous year's levels.

TABLE 4.5. *Rate of growth*[a] *of metals consumption in China and the world, 1981–1991* (%)

Major metal	China	Japan	Korea, Rep. of	USA	World
Aluminium (primary)	3.6	4.4	14.0	(0.6)	1.7
Cadmium	6.4	15.3	10.3	(2.9)	1.4
Copper (refined)	4.8	2.6	9.4	(0.7)	1.0
Lead (refined)	3.9	1.0	11.2	1.0	0.3
Magnesium	4.5	4.2	15.6	0.0	1.7
Nickel	4.9	6.2	22.4	(1.0)	2.2
Steel, crude[b]	6.1	3.4	13.4	(3.1)	0.4
Tin	5.4	1.3	14.3	(1.3)	0.5
Zinc	6.7	1.9	14.3	(0.6)	1.2

Notes: Figures are rounded; figures in parentheses are negative values.
[a] Average annual growth rates.
[b] Based on apparent consumption figures.

Sources: Metallgesellschaft (1992); World Bureau of Metal Statistics (1992); International Iron and Steel Institute (1993); Shen (1988).

To illustrate how China's recent consumption patterns compare with Japan, the Republic of Korea, the United States, and the entire world, Table 4.5 lists the average annual growth rates for nine metals for the 1981 to 1991 period. The figures clearly reveal that over the ten-year period analysed consumption of most metals grew significantly faster in China than in the world as a whole, the United States, and Japan. This is especially impressive considering the large size of China's metals consumption. The most notable growth rates in consumption are for crude steel, zinc, cadmium, tin, nickel, copper and magnesium—all of which were consumed at average annual growth rates at or above 4.5 per cent.

Of the countries listed in Table 4.5, only the Republic of Korea exhibited stronger average annual rates of growth in consumption for a majority of the metals (including crude steel) as compared to China. Korea, classified as one of Asia's newly industrializing economies (NIEs), has undergone rapid economic growth since 1986. The nation's GDP rose by 8.3 per cent in 1991, following a 9 per cent increase in 1990. Though the growth rates of metals consumption in the Republic of Korea are generally stronger than in China, the amounts of commodity consumed in China are significantly greater than those in Korea. As was shown in Table 4.4, only nickel was consumed by Korea in an amount nearly equal to that in China. Owing to the large volumes, China's recent growth in metals consumption has noticeably influenced world market conditions.

As with metals, energy use is growing rapidly in China in concurrence with modernization. Today China is the third largest energy-consuming nation in the world. By 2000, oil consumption is expected to be between two and three times higher than the 1980 level, coal use just more than two times greater, and natural gas consumption three times higher. Between 1985 and 1990, oil demand growth in China averaged 6.7 per cent per year (Fridley 1991*a*), although this aggregate figure does not accurately reflect high growth rates for petrochemical feedstocks, gasoline, and diesel (11.5, 10, and 8.5 per cent respectively), a decline in the consumption of kerosene, and slow growth in fuel oil demand (4.5 per cent). In the 1990s, demand growth is expected to vary between 5 and 7 per cent, reaching 2.7–2.8m. barrels per day (mmb/d) in 1995 and 3.4–4.0 mmb/d in 2000 (Fridley 1991*a*).

As is highly recognized, coal consumption in China is extraordinarily large, at 543.8m. tonnes oil equivalent in 1991 (Table 4.6). This compares to 477.9m. tonnes oil equivalent in the United States, and less than 79m. tonnes oil equivalent in energy-deficient Japan. Consumption of coal in China accounted for more than 25 per cent of the world's demand in 1991, with the United States being second with a share of 21.9 per cent.

The Chinese economy will continue to rely on coal as its principal energy source to the end of the century and beyond. Continuing rapid electrification in China will keep power demand growing at between 8 and 10 per cent annually during the 1990s, resulting in a near doubling of electricity demand by 2000. According to Johnson (1992), coal consumption in China will reach 1,335m. tonnes in 2000, and 1,560m. tonnes in 2010. This implies an increase in coal consumption of almost 500m. tonnes during the period 1990–2010. Chinese government officials are therefore seeking ways to improve the performance and efficiency of existing power facilities.

TABLE 4.6. *Energy consumption in China and the world, 1991*[a] (m. tonnes unless otherwise noted)

Commodity	China	Japan	Australia	USA	World
Crude oil	117.9 (3.8)	247.3 (7.9)	31.1 (1.0)	774.4 (24.7)	3,141.4
Coal[b]	543.8 (24.9)	78.7 (3.6)	39.9 (1.8)	477.9 (21.9)	2,186.2
Natural gas[b]	13.4 (0.8)	49.5 (2.8)	16.5 (0.9)	507.7 (28.7)	1,769.7
TOTAL PRIMARY ENERGY[b,c]	685.7 (8.7)	438.2 (5.6)	88.8 (1.1)	1,952.0 (25.0)	7,807.6

Note: Figures are rounded.

[a] Figures in parentheses represent country's % share of total world consumption.

[b] Million tonnes oil equivalent.

[c] Includes crude oil, coal, natural gas, hydroelectric power, and nuclear energy. Primary energy comprises commercially traded fuels only. Excluded therefore are fuels such as wood, peat, and animal waste, which though important in many countries are unreliably documented in terms of consumption statistics.

Source: British Petroleum Company (1984–92 (June 1992)).

China's primary energy needs are largely satisfied by coal, which accounted for nearly 80 per cent of supplies to the nation in 1991. Although high, this share of the total energy mix has dropped from the early 1950s, when coal supplied around 95 per cent of final consumption, but it is up from a low of 70 per cent in 1976. Crude oil, which represented 17.2 per cent of China's total primary energy consumption in 1991, has grown in importance as an additional source of energy in China since the late 1950s/early 1960s. It was during these years that several oil discoveries were made in China, and crude was found to be a relatively inexpensive fuel needed to supplement coal. Increased usage of oil has been encouraged in China owing to its nearness to industrial centres in the north-east—implying an easing of transportation bottlenecks—and the requirements of crude as petrochemical feedstocks. None the less, the abundance of coal in China and the relatively slow development of offshore and additional onshore oil deposits suggest that coal will remain as the critical energy source in China for the foreseeable future.

Table 4.7 compares the rates of growth of energy consumption in China, Japan, Australia, and the United States with the world for the 1981 to 1991 period. Australia represents the third largest energy-consuming country in the Pacific Basin behind China and Japan. Energy consumption in China grew steadily and rapidly during the 1980s, with rates of usage of crude oil, coal, and natural gas greatly exceeding world averages. With China's total primary energy consumption averaging 5.3 per cent growth per year since 1981, demands on domestic energy supplies have risen tremendously. Natural gas output, although still insignificant, has steadily increased in recent years owing to growing energy requirements. Mostly concentrated in Sichuan province, production of natural gas has helped fuel the industrialization of the Upper Yangtze region.

TABLE /. *Rate of growth*[a] *of energy consumption in China and the world, 1981–1991* (%)

Commodity	China	Japan	Australia	USA	World
Crude oil	3.4	1.0	0.6	0.4	0.8
Coal	5.9	2.2	2.8	1.8	1.8
Natural gas	2.6	7.4	4.5	0.2	3.0
TOTAL PRIMARY ENERGY[b]	5.3	2.6	2.2	1.1	2.0

Note: Figures are rounded.

[a] Average annual growth rate.

[b] Includes crude oil, coal, natural gas, hydroelectric power, and nuclear energy. Primary energy comprises commercially traded fuels only. Excluded therefore are fuels such as wood, peat, and animal waste, which though important in many countries are unreliably documented in terms of consumption statistics.

Source: British Petroleum Company (1984–92 (June 1992)).

With output of crude oil from China's major onshore oil fields (Daqing, Shengli, and Liaohe) still remaining at near peak or peak levels despite earlier dire predictions, demand for oil in China is expected soon to outstrip domestic supplies. This would return China to being a net importer of crude oil for the first time since the 1950s and early 1960s. As for coal resources, supply and demand are expected to remain in rough balance through the 1990s, although a recent slow-down in the growth of output may presage a shortfall by the end of the century.

CONCLUSIONS: A HUGE BASE AND INDUSTRY

Varied geologic conditions across China have dictated the distribution patterns of mineral and energy resources. At least 160 types of resources have been identified in the country. Some of China's reserves are among the largest in the world, including antimony, barite, coal, magnesite, molybdenum, pyrite, rare earths, tin, and tungsten. Yet China's abundant mineral and energy resources have yet to be fully identified and exploited.

With a population of more than 1.1 billion, China is a major minerals- and energy-consuming nation. In addition, with vast reserves of metals, non-metals, and energy resources, the country ranks among the world's leaders in mining output. An extraordinarily huge mining industry consisting of a multitude of ministries, corporations, and research organizations supervises all mineral-related activities. Many of these administrative bodies operate at three different levels of authority, namely the central government, provincial, and local ranks. Decentralization during the 1980s and 1990s has helped to shift some of the powers of the central ministerial-level enterprises to their provincial and local branches as well as to non-state operations, which now account for nearly half of China's gross industrial output.

NOTES

1. It is often reported in geologic literature that mineral deposits in China are, on average, of lower grade or quality than similar deposit types worldwide. However, this generalization is disputed by some geologists. Based on statistical averages and the knowledge that China's land area is extremely large, mineral deposits in China are most likely to be of the same average grade as those found elsewhere in the world. It is, perhaps, because of differences in the manner in which Chinese geologists define their reserves that make it appear as if average grades of deposits in China are lower than similar deposits

worldwide. In general, the marginal cut-off grade of deposits in China includes all technologically and economically recoverable ore; in many Western countries, however, the marginal cut-off grade is that portion of the ore body that is economically extractable.

2. China completed its first ever systematic nationwide minerals and energy survey in 1985. A sample, experimental and small-scale study was also conducted in 1984 and 1980. The 1985 nationwide survey was supervised by the state and involved such government organs as the Ministry of Geology and Mineral Resources. The survey examined the number of mines and mineral and energy resources discovered in China, and production, consumption, and employment levels in the nation.
3. For a comprehensive listing of major geological studies completed in China between 1953 and 1984, see Ministry of Geology and Mineral Resources (1982; 1983; 1984).
4. Mineral reserves in China are classified according to the 'Tentative Criteria for Mineral Reserves Classification (TCCR)' which was issued by the National Mineral Reserves Committee in 1959. According to the classification scheme, mineral reserves in China are categorized as 'industrial' and 'non-industrial', or 'prospective'. Industrial reserves are technologically recoverable or mineable, with little consideration of economics. In contrast, non-industrial reserves are not technologically recoverable for industry.

   The TCCR classification scheme divides reserves into four types and six grades: Type 1—reserves subject to mining, generally A_1 grade; Type 2—reserves subject to designing, generally A_2, B, and C_1 grades; Type 3—prospective reserves, of C_2 and D grades; and Type 4—geological reserves, of D grade. Industrial reserves are of grades A, B, C_1, and C_2; non-industrial or prospective reserves are of grade D. According to the TCCR, reserve amounts of grades A_1, B, and C_1 are subject to as much as ±20%, ±30%, and ±45%, respectively.
5. The mine has estimated reserves of 36m. tonnes, or approximately 70% of the world's total. Tests indicate that 97% of the deposit belongs to the light rare earth variety, containing 25% lanthanum, 50% cerium, and 15% neodymium, in addition to some medium-variety rare earth metals including europium and samarium (Xu 1992). Rare earths were first discovered at Bayan Obo in 1934, when geologists found bastnasite and monazite.
6. For lengthy discussions on China's government organization and responsibilities, readers are referred to Xue Muqiao (1982) and Saich (1981).
7. China's government organizational reforms completed in Sept. 1988 included the following changes: (1) the former State Planning Commission and State Economic Commission became merged into a new State Planning Commission; (2) the former Ministry of Labour and Personnel has become two separate ministries, the Ministry of Labour and the Ministry of Personnel; (3) the former State Raw Materials Bureau has become the new Ministry of Materials; (4) the former Ministry of Water Resources and Electric Power has been divided up into the new Ministry of Water Resources and new Ministry of Energy; (5) the former Ministries of Coal, Petroleum, and Nuclear Industries, as well as part of the former Ministry of Water Resources and Electric Power, have been merged into the new Ministry of Energy; (6) the former urban and rural department of the Ministry of Urban and Rural Construction and Environmental Protection has become the new Ministry of Construction; (7) the former Ministry of Electronics Industry and State Machine-Building Industry Commission have become the new Ministry of Machine-Building and Electronics Industry; (8) the former Ministries of aviation industry and space industry to have been merged into the new Ministry of

Aeronautics and Astronautics Industry; (9) the former Ministry of Agriculture, Animal Husbandry and Fishery has been newly named the Ministry of Agriculture; and (10) the Xinhua News Agency, formerly under the State Council, is now independent as a result of these reforms. Further organization reform of China's government is likely in the 1990s.

8. Based in large part on discussions between 1986 and 1992 with Chinese officials from the Ministries of Geology and Mineral Resources, Metallurgical Industry, and Energy, the Institute of Geology (Academia Sinica), the Wuhan College of Geology (Hubei province), and CNNC.
9. Using linear regression modelling, the Technical Information Centre of the Ministry of Metallurgical Industry recently forecast China's steel products demand by the year 2000. Based on past rates of economic growth and steel products use intensity, the models predicted that China's consumption of steel goods in 2000 would range between 124m. and 147m. tonnes. With the adoption of new steel-saving technologies and substitition practices, however, actual steel products consumption in 2000 is predicted to be between 118m. and 123m. tonnes. The forecast value of steel products demand exceeds the forecast value of steel products output in 2000, though a relatively large amount of steel products can be saved by advanced science and technology and materials substitution. The forecasting models and results are presented in Shen (1988).

   In 1991, colleagues in the Economic Research Centre of the University of Western Australia generated a vector autoregression (VAR) model of the Chinese economy to forecast steel demand to the year 2000. A VAR model consists of a system of equations in which each variable is regressed on its own lagged values and on the lagged values of all the other variables in the model. An important advantage of VARs is that there is no need to specify exogenous variables over the forecast period. The model, described in Chen *et al.* (1991), predicted that Chinese steel demand would grow to close to 100m. tonnes by 2000.

5
Commodity Trade and International Co-operation

PRIORITIES, OVERALL TRADE

China employs foreign trade as a means of advancing development goals and achieving political objectives. The composition of China's commodity exports and imports therefore changes with shifts in priorities and economic strategies.

The implementation of agricultural and industrial reforms in 1979 resulted in more than a doubling of economic output. During the 1980s China's economy expanded at a relatively strong pace, averaging nearly 10 per cent growth per annum in real terms. Vigorous economic growth has continued into the 1990s. This expansion has led to an increased consumption of energy, raw materials, and finished goods within China, with an increasing volume of these materials being imported from foreign sources. Chinese demand for imports is very large, given the shortages and bottlenecks in China's economy. As China's foreign trade rules and regulations were relaxed with reform, imports have tended to grow much more strongly than exports. By the early 1990s, however, the value of China's exports outpaced imports, providing the country with a net trade surplus.

Between 1980 and 1990, China's overall exports and imports grew faster than those of any other economy in the Asia–Pacific region. The growth in trade did not slow down in 1991, when total exports and imports were valued at $135.7bn., exhibiting an increase of 17.6 per cent over 1990. Since 1990, China's balance of foreign trade has been positive, as the value of Chinese exports continues to exceed that of its imports. In 1991, China's overall trade surplus stood at an estimated $8.1bn. (Table 5.1), slightly less than the previous year. During that year, the total volume of foreign trade approached $136bn., of which $71.9bn. was attri-buted to exports and $63.8bn. to imports. From the previous year, the value of Chinese exports rose briskly at 15.8 per cent in US dollar terms, yet the import value grew at an even faster pace, by 19.5 per cent.

China's overall trade in 1991 was characterized by a strengthening of reform in the foreign trade system, an improved export mix, and increased barter trade. While China's barter trade with the former Soviet Union and Eastern Europe increased slightly, new links were established with several developing nations for sales of electronic goods and machinery. Export growth in 1991 can be attributed to continued reform in China's foreign trade system, the contract responsibility system by local governments and enterprises, and encouragement provided to exporters. The sustained strong export growth in 1991 was particularly notewor-

TABLE 5.1. *China's overall trade, 1991* (US$bn.)

	1991	1990	Change (%)
Total value of trade	135.7	115.4	17.6
Total value of exports (f.o.b.)	71.9	62.1	15.8
Total value of imports (c.i.f.)	63.8	53.4	19.5
BALANCE (+ export surplus − import surplus)	+8.1	+8.7	Worsening

Note: Figures are rounded.
Source: General Administration of Customs, PRC (1990–2).

thy given the worldwide recession and the removal of export subsidies in China, often considered the source of competitiveness of Chinese products in global markets. Excessive import growth in 1991 resulted from the continued expansion of the Chinese economy and industrial base and a large-scale increase in the purchase of foreign consumer goods. Exports of manufactured products increased in 1991, particularly in the fields of chemical products, textiles, and industrial goods. Manufactured goods make up more than two-thirds of China's total export value.

With the objective of quadrupling China's 1980 GVAIO by the year 2000 still intact, the nation has become increasingly reliant on foreign trade to facilitate economic growth. Total foreign trade reportedly reached $165bn. in 1992, a growth of more than 20 per cent. New international markets have been sought for China's products to help finance economic expansion while, at the same time, energy, raw material, and industrial goods have been imported in greater volumes to satisfy domestic demand. Owing to the importance of minerals and energy to China's economic development, minerals and mineral products, mining technology and equipment, and energy represent an important component of the nation's overall foreign trade. In recent years China's minerals industry has striven to boost output of semi-finished to finished value-added goods for not only domestic consumption but for export. The additional foreign exchange acquired through export of value-added goods is becoming an increasingly important source of revenue in the light of China's metals trade deficit and overall currency requirements.

MINERALS AND ENERGY TRADE

Minerals and energy trade in China achieves several objectives including, for exports, greater foreign exchange earnings, and government industrial revenues, and for imports, the acquisition of necessary raw materials for domestic industrial utilization. Other benefits via trade include technology transfers, improved quality of raw material inputs, an increased number of import (and export) sources (and destinations), and improved employment opportunities. Minerals and energy trade in China involves several key organizations at the national level, primarily the many subsidiaries of MOFERT, including the China National Chemicals Import and Export Corporation (SINOCHEM) and the China National Metals and Minerals Import and Export Corporation. Additional major trading branches belong to other ministries and departments, including the China National Offshore Oil Corporation, China National Petrochemical Corporation (SINOPEC), China National Metallurgical Import and Export Corporation, China National Nonferrous Metals Import and Export Corporation, and the Economic Relations and Trading Company (of the China National Nonmetallic Minerals Industry Corporation).

Table 5.2 provides a value breakdown by commodity group of China's mineral and energy exports between 1970 and 1991. The commodity grouping is based on the United Nations' Standard International Trade Classification (SITC) Revision 2 classification scheme, where nine mineral and energy groups are included, namely crude fertilizers and minerals (D27), metalliferous ores and scrap metal (D28), coal, coke, and briquettes (D32), petroleum and petroleum products (D33), gas, natural and manufactured (D34), non-metallic mineral manufactures (D66), iron and steel (D67), non-ferrous metals (D68), and manufactures of metals (D69). A detailed listing of the products within each of these SITC Revision classification system categories can be found in United Nations Statistical Office (1975).

Mineral and energy exports from China in 1991 were worth an estimated $11.08bn., or 15.4 per cent of the country's overall export value of $71.9bn. The value of petroleum and petroleum product exports far exceeds that of any other mineral or energy commodity group. In 1991 China's petroleum and petroleum product exports were valued at nearly $4bn., or more than 35 per cent of the total value of all mineral and energy exports that year. China's exports of iron and steel goods were second in value behind petroleum and petroleum products, accounting for just more than 15 per cent of the total. The iron and steel category includes such items as pig iron, ferroalloys, ingots of iron and steel, iron and steel bars, rods, angles, shapes, sheets and plates, and castings.

It was not until the 1970s that China's petroleum and petroleum product exports grew substantially in both quantity and value. In 1970 the share of petroleum and petroleum product export value represented just under 3 per cent of the total value of mineral and energy exports, whereas by 1975 they accounted for a remarkable 51.6 per cent. In terms of tonnage, between 1970 and 1975 the quantity of refined petroleum products exported abroad rose more than tenfold, while the amount of crude oil sold abroad increased more than 50 times.

Despite the huge size of China's metals and non-metals industrial base, until recently the value of non-fuel mineral exports from the country was relatively small. Table 5.2 reveals that the total value of mineral exports has risen significantly since 1970, at an average annual growth rate of 17.4 per cent in current terms.

TABLE 5.2. *China's mineral and energy exports, 1970–1991*[a] ($US '000)

Mineral group[b]	1970	1975	1980	1985	1989	1990	1991
Crude fertilizers and minerals	42,000	113,000	259,000	358,000	543,720	515,750	511,520
Metalliferous ores and scrap metal	27,400	55,800	168,000	246,000	385,610	212,060	154,830
Coal, coke, and briquettes	50,000	159,000	242,000	579,000	680,370	754,980	829,380
Petroleum and petroleum products	7,460	862,000	4,120,000	6,630,000	3,581,450	4,459,940	3,975,440
Gas, natural and manufactured	0	0	1,030	4,700	1,720	2,610	3,580
Non-metallic mineral manufactures	59,800	172,000	401,000	398,000	792,510	1,316,220	1,668,320
Iron and steel	21,700	71,700	267,000	168,000	708,890	1,282,540	1,669,120
Non-ferrous metals	23,300	113,000	187,000	305,000	453,720	597,280	560,610
Manufactures of metals	39,300	119,000	408,000	570,000	1,209,640	1,437,330	1,703,680
Total metals and non-metals	214,000	646,000	1,690,000	2,040,000	4,094,090	5,361,180	6,268,080
Total energy	57,400	1,020,000	4,360,000	7,210,000	4,264,330	5,217,530	4,808,400
TOTAL	271,000	1,670,000	6,050,000	9,250,000	8,358,420	10,578,710	11,076,480

Note: Figures are rounded.

[a] f.o.b. basis.

[b] SITC categories D27, D28, D32, D33, D34, D66, D67, D68, and D69, respectively.

Sources: General Administration of Customs, PRC (1983–92); United Nations (1984–90); Minerals Policy Program (1993).

China's mineral exports amounted to $6.3bn. in 1991, which was more than 29 times the 1970 value. Values of mineral exports are fairly evenly distributed through four of the six commodity groupings, with the metalliferous ores and scrap metals group accounting for the lowest share of the total value—2.5 per cent. The most valuable export groups, manufactures of metals, accounted for 27.2 per cent of the total export value, with iron and steel accounting for almost the same, 26.6 per cent. Exports of non-ferrous metals have risen tremendously since 1980, as evidenced by the boost in value. Almost all of China's global trade of non-ferrous metals is handled by the China National Nonferrous Metals Import and Export Corporation (CNIEC)—the trading arm of CNNC.

Figure 5.1 illustrates the value (f.o.b.) of China's mineral and energy exports during the years 1987, 1989, and 1991. Such a time span provides a useful indication of the impacts of more than a decade of reform. The figure permits easy comparison of the values of each commodity grouping for the five-year period. As is seen in Table 5.2, petroleum and petroleum products were by far the most valuable export group in 1991, overwhelmingly large when compared to the other commodity groupings. To better reveal the relative contribution to export value of the other eight commodity groups, Figure 5.2 depicts China's 1987 to 1991 export values exclusive of petroleum goods. Here, manufactures of metals and iron and steel were the second and third most valuable export groups in 1991, combined accounting for more than 30 per cent of total mineral and energy export value.

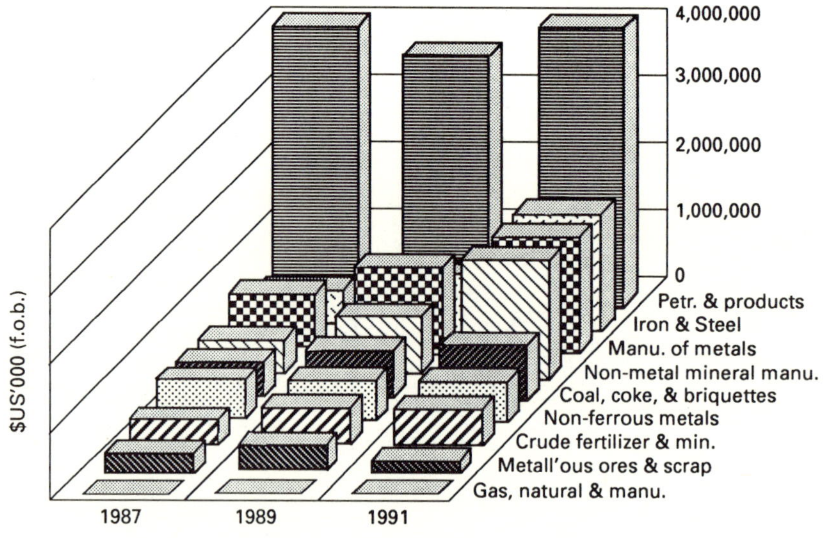

FIG. 5.1. Mineral and energy export values of China, 1987–1991

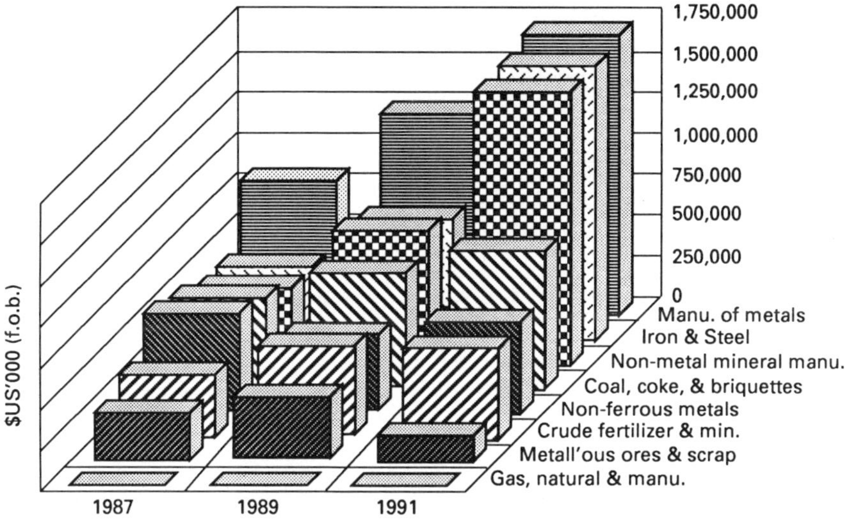

FIG. 5.2. Mineral and energy export values of China (exclusive of petroleum goods), 1987–1991

Owing to its abundance of coal production, China does not yet require substantial imports of energy commodities, although foreign suppliers are accounting for an increasingly large share of the nation's fuel requirements. Table 5.3 lists 1970–91 values of energy as well as mineral imports in thousand US dollars. As a group, energy imports accounted for 25 per cent of the total mineral and energy import value in 1991, compared to less than 10 per cent just three years prior. Petroleum and petroleum products, the major energy import, represented 95.1 per cent of the total energy import value in 1991. China's rapid economic growth in the 1980s, coupled with conservation efforts, caused a gradual shift in petroleum product demand. Until 1980, nearly half of China's petroleum products were burned, while output of motor fuels stagnated, with low growth in production of automotive, tractor, and transport vehicles (Fridley 1988). After 1980, however, changing internal and external economic conditions, including deregulation of agriculture, led to increased demand for transport fuels in China that exceeded supplies. Thus, petroleum product imports have been steadily rising in China, despite recent progress in upgrading the nation's refining system.

Values of China's mineral imports are not distributed evenly across commodity groupings as are the exports. The most valuable category of imports is iron and steel, which represented just under 47 per cent of China's total mineral import value in 1991 (Table 5.3). Historically, China has been deficient in domestic iron and steel production and has had to import the materials in large quantities. The nation's domestic capability of producing enough iron and steel (of sufficient

TABLE 5.3. *China's mineral and energy imports, 1970–1991*[a] ($US '000)

Mineral group[b]	1970	1975	1980	1985	1989	1990	1991
Crude fertilizer and minerals	11,000	14,500	74,500	85,400	48,530	33,540	53,820
Metalliferous ores and scrap metal	19,700	73,600	136,000	290,000	788,310	950,290	1,220,320
Coal, coke, and briquettes	29,000	86,200	124,000	132,000	90,920	74,170	46,260
Petroleum and petroleum products	11,200	40,500	140,000	308,000	1,465,240	1,053,840	1,846,560
Gas, natural and manufactured	0	432	5	1,500	15,670	27,190	48,550
Non-metallic mineral manufactures	28,000	28,700	154,000	353,000	520,450	453,150	442,960
Iron and steel	389,000	1,470,000	2,250,000	5,870,000	5,796,530	2,852,030	2,694,180
Non-ferrous metals	199,000	404,000	444,000	1,320,000	1,114,280	579,070	815,640
Manufactures of metals	18,200	117,000	368,000	481,000	627,780	539,640	538,760
Total metals and non-metals	664,000	2,100,000	3,430,000	8,390,000	8,895,930	5,407,720	5,765,680
Total energy	40,200	127,000	264,000	442,000	1,571,830	1,155,200	1,942,370
TOTAL	704,000	2,230,000	3,690,000	8,830,000	10,467,760	6,562,920	7,708,050

Note: Figures are rounded.

[a] c.i.f. basis.

[b] SITC categories D27, D28, D32, D33, D34, D66, D67, D68, and D69, respectively.

Sources: General Administration of Customs, PRC (1983–92); United Nations (1984–90); Minerals Policy Program (1993).

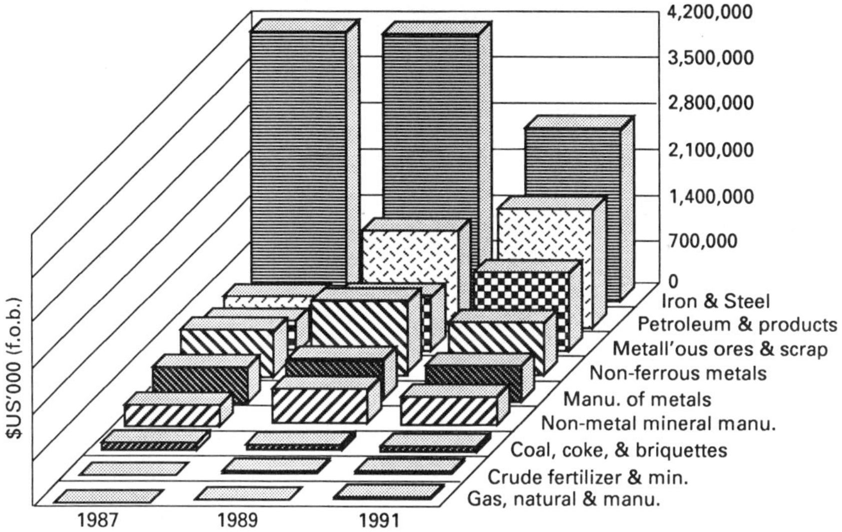

FIG. 5.3. Mineral and energy import values of China, 1987–1991

quality) to enhance and sustain modernization is limited. Since 1970, iron and steel imports have accounted for 58.6 (1970), 70.0 (1975), 65.6 (1980), and 70.0 (1985) per cent of the nation's total metals and non-metals import value. Like Figure 5.1, Figure 5.3 graphically depicts China's mineral and energy import values by

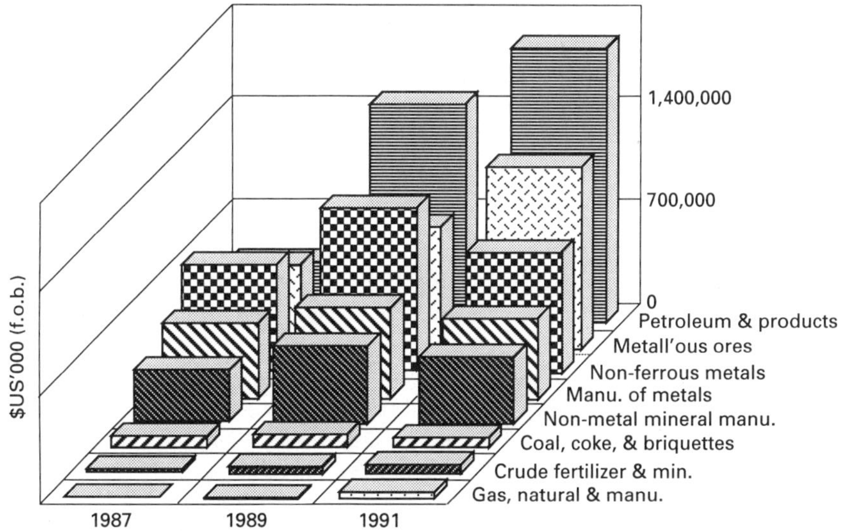

FIG. 5.4. Mineral and energy import values of China (exclusive of iron and steel), 1987–1991

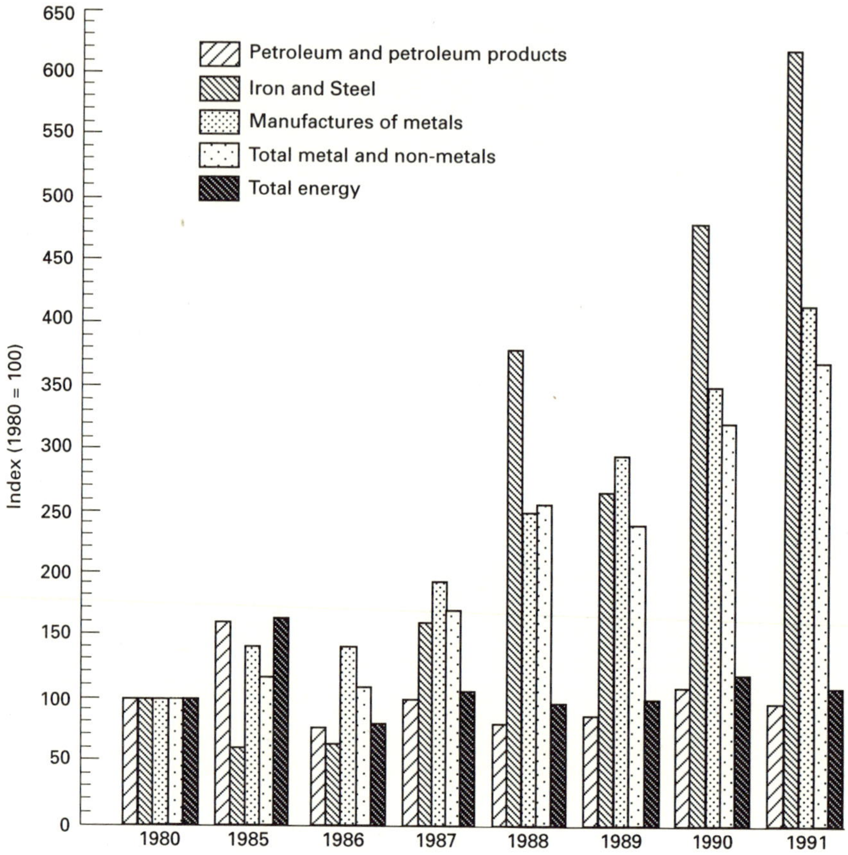

FIG. 5.5. Indices of mineral and energy export values of China, 1980–1991

commodity group for the years 1987, 1989, and 1991. As is shown, the value of imported iron and steel is high relative to that of other groups, although not as overwhelmingly high as in previous years owing to strong efforts to curb imports. After iron and steel, petroleum and petroleum products and metalliferous ores and scrap are imported in significant quantities. The metalliferous ores and scrap category includes iron ore and concentrates, waste and scrap metal of iron, steel, copper, nickel, aluminium, lead, zinc, and tin, and ores and concentrates of base metals, uranium, non-ferrous metals, and precious metals. Figure 5.4 illustrates the relative contributions to import value of eight commodity groups, excluding iron and steel.

With the advancement of economic reform during the 1980s and 1990s, fluctuating oil prices, and great strides in modernization, China's minerals and energy trade in recent years has been particularly volatile. Worries about an overheated economy and lack of foreign exchange caused Chinese officials to alter some import and export patterns in late 1988, with the introduction of an economic austerity programme. To illustrate some of these changes over a period of years since 1980, Figures 5.5 and 5.6 display trends of Chinese exports and imports of selected commodity groups. Trends are based on indexed trade values with 1980 serving as a base year. Those groups illustrated are the three most valuable export or import categories in 1991. Changes in indexed values of all mineral and metal exports/imports as well as all energy exports/imports are also illustrated in the figures.

Among the most notable trends illustrated in Figure 5.5 is the apparent peak in export value of petroleum and petroleum products in 1985. After a steady rise in export value beginning in 1983, the drop in world oil prices severely reduced China's petroleum export earnings in 1986 and beyond. Many analysts believe that the value of petroleum and petroleum product exports from China will never exceed that in 1985 owing to dramatically increasing domestic demand. The slight but noticeable increase in petroleum and petroleum export value in 1990 was caused by a deliberate increase in exports by China during that year to offset reduced world supplies due to the Persian Gulf War crisis.

Other important trade patterns evident in Figure 5.5 include the dramatic growth in iron and steel export value between 1987 and 1988 and the steady rise in export value of all mineral and metal commodities since 1986. Much of the growth in exports of iron and steel goods in 1990 and again in 1991 was shipped to Hong Kong (for subsequent re-export) and Japan. Combined, both countries accounted for more than 44 per cent of China's iron and steel exports, with Japanese imports rising nearly 39 per cent between 1990 and 1991. The 1989 ban on the exports of selected non-ferrous metals from China clearly slowed the increase in export value of non-ferrous metals begun in 1986, but had relatively little impact on export growth of all minerals and metals.

As for imports, Figure 5.6 illustrates an alarming and dramatic rise in the import value of energy commodities beginning in 1985. Petroleum and petroleum products imports in 1991 were valued at $1.8bn. or, remarkably, six times the 1985 import value. Currently, China still exports more oil than it imports, yet the situation may soon reverse. Fridley (1991*b*) concludes that China is likely to become a net oil importer, after examining various assumptions on production, refinery expansion and demand. In one of his scenarios of higher demand and lower production, China will depend on importing refined products for up to 1.7m. b/d and crude oil for about 380,000 b/d in the year 2000. Figure 5.6 also shows a strong rise in the import of metalliferous ores and scrap between 1985 and 1991.

Despite slightly lower world prices, high levels of energy exports (particularly petroleum and petroleum products) allowed China's overall balance of trade with respect to minerals and energy to remain positive in 1991. Table 5.4 contains values in US dollars of China's mineral and energy trade deficits and surpluses since

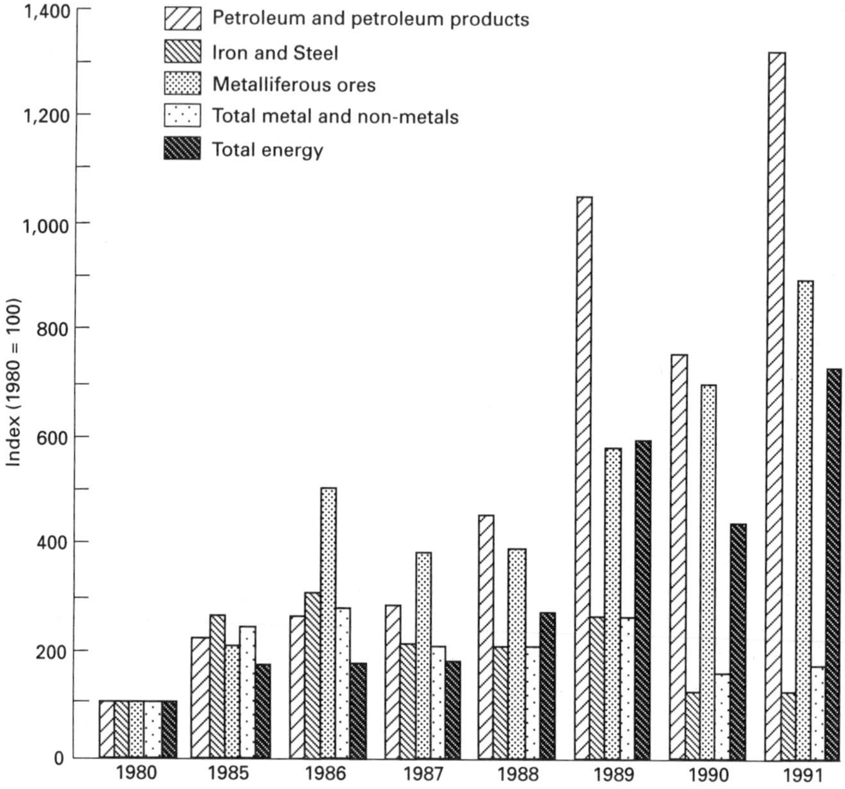

FIG. 5.6. Indices of mineral and energy import values of China, 1980–1991

1970. As shown, the surplus value of energy trade exceeded the deficit value of minerals and metals trade by $3.37bn. Such a surplus was not available as recently as 1989, however, when a more than doubling of petroleum and petroleum product imports occurred resulting in an overall deficit of greater than $2.1bn. Given the significance of petroleum and petroleum product imports to China's overall mineral and energy trade balance, continued growth in imports may jeopardize China's ability to maintain a positive trade balance.

The overall trade situation has been rather erratic in China since 1970, as the mineral and energy industries suffered trade deficits in the early 1970s, followed by trade surpluses in the early 1980s, which again were followed by deficits towards the mid-1980s. Many factors contributed to this behaviour, including: (1) dramatic changes in import values of iron and steel; (2) a remarkable rise in the export value of petroleum and petroleum products until 1985 when declining oil prices and growing domestic consumption curtailed further increases; and (3) a strong in-

TABLE 5.4. *China's mineral and energy trade surplus/deficits,*[a] *1970–1991* ($US '000)

Mineral group[b]	1970	1975	1980	1985	1989	1990	1991
Crude fertilizer and minerals	31,000	98,500	184,500	272,600	495,190	482,210	457,700
Metalliferous ores and scrap metal	7,700	−17,800	32,000	−44,000	−402,700	−738,230	−1,065,490
Coal, coke, and briquettes	21,000	72,800	118,000	447,000	589,450	680,810	782,120
Petroleum and petroleum products	−3,740	821,500	3,980,000	6,322,000	2,116,210	3,406,100	2,128,880
Gas, natural and manufactured	0	−432	1,025	3,200	−13,950	−24,580	−44,970
Non-metallic mineral manufactures	31,800	143,300	247,000	45,000	272,060	863,070	1,225,360
Iron and steel	−367,300	−1,398,300	−1,983,000	−5,702,000	−5,087,690	−1,569,490	−1,025,060
Non-ferrous metals	−175,700	−291,000	−257,000	−1,015,000	−660,560	18,210	−255,030
Manufactures of metals	21,100	2,000	40,000	89,000	581,860	897,690	1,164,920
Total metals and non-metals	−450,000	−1,450,000	−1,740,000	−6,350,000	−4,801,840	−46,540	502,400
Total energy	17,200	893,000	4,100,000	6,770,000	2,691,710	4,062,330	2,866,030
TOTAL	−433,000	−557,000	2,360,000	418,000	−2,110,130	4,015,790	3,368,430

Note: Figures are rounded.

[a] Trade surplus/deficits calculated as f.o.b. exports minus c.i.f. imports to reflect actual payments and costs to China.
[b] SITC categories D27, D28, D32, D33, D34, D66, D67, D68, and D69, respectively.

Source: Tables 5.2 and 5.3 above.

crease in the import value of non-ferrous metals until 1985. China's minerals trade deficit reached a peak in 1986, when the deficit reached an estimated $4.4bn. Although the deficit worsened steadily in the 1980s, a marked drop in iron and steel and non-ferrous metal imports in 1990 helped transform a negative trade balance into a surplus during that year. Since 1990, growing imports of metalliferous ores and scrap metal, non-ferrous metals, and petroleum and petroleum products has caused concern among Chinese government officials about an overheated economy.

COMMODITY BY COMMODITY TRADE FLOWS

Tables 5.5 and 5.6 disaggregate China's energy and minerals trade (both exports and imports) on a commodity-specific basis. The commodities are listed in order of value for 1991.

China's energy exports and imports are dominated by crude petroleum and refined petroleum products. The value of crude oil plus refined petroleum exports in 1991 accounted for nearly 79 per cent of that year's total energy export value. Disaggregated into components, crude oil exports accounted for $3bn., or 61.5 per cent of total energy exports, while refined petroleum product exports contributed 17.1 per cent of 1991 export value, or $823m. Coal accounted for much of the remaining export value that year. Crude petroleum and refined petroleum products accounted for most of China's imported energy commodities in 1991, at a cost of $926m. for crude oil and $902m. for refined products. In just one year, the value of imported crude oil to China rose 118.5 per cent, from $423.8m. in 1990.

In terms of tonnage, China exported slightly more than 27m. tonnes of crude oil and refined petroleum products in 1991, up 8.3 per cent from the preceding year. China relies heavily on petroleum and petroleum product exports for the acquisition of foreign exchange. Such earnings are used to purchase foreign goods and equipment, which often include energy-consuming products such as motor vehicles. As a consequence, China follows a practice of exporting part of its limited domestic energy supplies while exacerbating the situation by creating additional demand.

Unlike the energy commodities, China's metal and non-metal exports are more evenly distributed in value terms, with only iron and steel products, cement, and ferrosilicon representing significant shares of the total in 1991. The most valuable export commodity is iron and steel products, worth an estimated $875.3m. in 1991, or 14 per cent of total mineral export value. This represents a 25 per cent increase in value from 1990 and a nearly 200 per cent increase from 1988, reflecting China's overall ambition to increase production and exports of metals in order to earn needed foreign exchange. In terms of non-ferrous metals, copper products, and tin and its alloys were the most significant exports, with copper product exports in 1991 valued at $108.1m., nearly $21m. greater than exports of

TABLE 5.5. *China's energy trade by commodity, 1991*

EXPORTS				*IMPORTS*			
Commodity	Value (US$m.)	Ranking in value	% of total[a]	Commodity	Value ($USm.)	Ranking in value	% of total[a]
Petroleum oil, crude	2,956.8	1	61.5	Petroleum oil, crude	926.4	1	47.7
Petroleum products, refined	822.9	2	17.1	Petroleum products, refined	901.9	2	46.4
Coal	748.5	3	15.6	Coal	47.2	3	2.4
Coke, semi-coke	80.4	4	1.7	Others	66.9	—	3.4
Others	199.8	—	4.2				

Note: Figures are rounded.

[a] Refers to total value of exports and imports of coal, coke, and briquettes (SITC classification D32), petroleum and petroleum products (D33), and gas, natural and manufactured (D34).

Sources: General Administration of Customs, PRC (1983–92); United Nations (1984–90); Minerals Policy Program (1993).

TABLE 5.6. *China's minerals trade by commodity, 1991*

EXPORTS Commodity	Value (US$m.)	Ranking in value	% of total[a]	*IMPORTS* Commodity	Value ($USm.)	Ranking in value	% of total[a]
Iron and steel products	875.3	1	14.0	Iron and steel products	2,534.6	1	44.0
Cement	442.9	2	7.1	Iron ore	565.9	2	9.8
Ferrosilicon	164.0	3	2.6	Copper and copper alloy, unwrought	259.3	3	4.5
Copper products	108.1	4	1.7	Aluminium products	216.0	4	3.7
Iron and steel wire	106.8	5	1.7	Copper products	177.8	5	3.1
Aluminium and its alloys, unwrought	104.0	6	1.7	Copper ore	154.6	6	2.7
Pig iron and cast iron	90.0	7	1.4	Chromium ore	71.7	7	1.2
Tin and tin alloy, unwrought	87.1	8	1.4	Aluminium and its alloys, unwrought	66.4	8	1.2
Fluorspar	65.8	9	1.0	Pig iron and cast iron	49.9	9	0.9
Aluminium products	55.7	10	0.9	Steel, billets, and rough forgings	30.0	10	0.5
Others	4,168.4	—	66.5	Others	1,639.5	—	28.4

Note: Figures are rounded.

[a] Refers to total value of exports and imports of crude fertilizers and minerals (SITC classification D27), metalliferous ores and scrap (D28), non-metallic mineral manufactures (D66), iron and steel (D67), non-ferrous metals (D68), and manufactures of metals (D69).

Sources: General Administration of Customs, PRC (1983–92); United Nations (1984–90); Minerals Policy Program (1993).

tin and tin alloys. Other significant export commodities include iron and steel wire, aluminium and its alloys, pig and cast iron, fluorspar, and aluminium products.

Although not among the top ten most valuable mineral exports listed in Table 5.6, an important export commodity group from China during the 1980s was the rare earth metals, including lanthanum, cerium, neodymium, gadolinium, europium, dysprosium, and yttrium. Rare earth metals are essential to the world's petroleum, aerospace, telecommunications, computer, and rapidly evolving superconductivity industries. China's reserves of rare earths are confirmed to be the largest in the world, estimated at 43m. tonnes or more than half of the world's supplies. China is at present the second largest producer of rare earth metals globally, edged out only slightly by the United States for that status in 1991. During that year, record levels of production, domestic consumption, and exports of rare earth metals occurred in China, indicating a robust industry. China exports high-purity rare earths and yttrium compounds, most notably to Japan and Europe.

Table 5.6 shows that on a commodity basis, the most valuable metals and non-metals imported into China in 1991 included iron and steel products, iron ore, copper and copper alloys, aluminium products, and copper products. Though China's policymakers realize that the country cannot be self-sufficient in mineral products, they are concerned about the growth of imports. In 1991, imports of most mineral commodity groups grew or remained steady as compared to the preceding year, with iron and steel remaining the most costly group imported to the nation. China is now a major importer of iron, steel, copper, and aluminium—the basic metals required by an industrializing society. It is also a substantial importer of some non-metals, including cement and fertilizer materials. These are of particular importance since they are raw material requirements of agriculture and heavy and light industry—the keystones of China's economic development. China remains a huge consuming nation with enormous potential as a marketplace as its people continue to strive for improved living standards.

CHINA'S TRADE PARTNERS

With the advent of economic reform and an opening of the Chinese economy, foreign trade has become an integral component in China's political relations to other countries. In the past, changing volumes of bilateral trade between a foreign country and China have usually indicated changing political relations between the two partners. Hence foreign trade has served as both a political linkage between foreign nations and China, and a means for China to acquire the necessary raw materials and products it lacks domestically.

As for overall trade, Hong Kong, Japan, the EC, the United States, and Germany were China's major trading partners in 1991. Owing to the significance of energy, metals, and non-metals in China's overall trade, many of these same countries/regions are leading trading partners of these commodities. China's trade with the Hong Kong and Macau regions in 1991 reached $140.5bn., accounting for 36.6 per cent of China's total trade.

Table 5.7 lists China's 1987, 1989, and 1991 trade partners for the three major energy commodity groups. As shown, significant quantities of petroleum exports have been sent to Japan during the 1980s and early 1990s, with much of it based on long-term sales contracts. Petroleum exports to Japan have been China's leading export product since the mid-1970s, contributing significantly to bilateral trade expansion between the two nations. Japan will likely remain China's biggest petroleum export customer in the years ahead, with Singapore the second largest. In recent years the role of Singapore as a conduit of China's petroleum and petroleum products to the Asia–Pacific region has grown. In 1991 Japan accounted for 48.4 per cent of China's total petroleum and petroleum product export revenues.

As with petroleum and its products, China's largest customer (although small in value) of coal exports is also Japan, where an estimated $249.8m. worth of coal was purchased in 1991. Japan annually imports about 4m. tonnes of coal from China. The EC and Hong Kong are China's other major coal export partners.

China's ability to export coal has expanded significantly in recent years, owing to improved infrastructure, additional coal production at the Pingshuo coal-mine (Shanxi province), and expanded selling efforts by the China National Coal Import and Export Corporation. The coal handling and transportation infrastructure in China has been greatly improved, as the rail line running between Datong and Beijing, and Beijing and the part of Qinhuangdao has been upgraded by doubling the track and electrification (Jia Yunzhen 1988*a*, *b*). China, which produced more than 1bn. tonnes of coal in 1991, expects to acquire larger shares of the global coal market in the years ahead. Orders for coal from East European countries are overwhelming Chinese training companies anxious to expand their exports. Eastern Europe still lacks hard currency for payments so many arrangements are being conducted through barter trade. While China's coal exports remained stable in 1991, the nation surpassed its goal of exporting 20m. tonnes.

The potential for China to increase coal exports to neighbouring Asia–Pacific countries is relatively large, as it is one of the lowest-cost producers of coal in the Pacific Rim and it contains some relatively good-quality coal fields. China's proximity to Japan, one of the largest coal importers in the region, is promising, although Japan's coal imports from China remain small in value today. Without the establishment of a coal trade agreement between Japan and China, there is little reason to believe that such exports will increase noticeably. Overall, if China's coal exports are to grow substantially in the near future, constraints like growing domestic demand and continued transport problems will have to be eased.

Table 5.8 lists China's 1987, 1989, and 1991 mineral trade destinations/sources by commodity grouping, country, and value. Five countries/regions received a

TABLE 5.7. *China's energy trading partners, 1987–1991*

	1987		1989		1991	
	Major countries	Values (US$m.)	Major countries	Values (US$m.)	Major countries	Values (US$m.)
EXPORTS						
Petroleum, petroleum products and related materials (D33)	Japan	2,076.5	Japan	1,788.7	Japan	1,932.2
	United States	1,833.4	Singapore	604.6	Singapore	713.9
	Singapore	630.0	United States	580.5	United States	678.0
Coal, coke, and briquettes (D32)	Japan	153.3	Japan	188.0	Japan	249.8
	North Korea	96.1	Hong Kong	79.9	EC	135.0
	Hong Kong	90.0	North Korea	72.8	Hong Kong	97.3
Gas, natural and manufactured (D34)	Hong Kong	1.2	Hong Kong	1.0	Hong Kong	1.2
	Macau	1.2	Macau	0.6	Macau	0.8
	United States	0.7	Singapore	Negl.	Japan	Negl.
IMPORTS						
Petroleum, petroleum products and related materials (D33)	Singapore	180.8	Singapore	756.6	Singapore	552.4
	Soviet Union[a]	68.7	Indonesia	193.5	Indonesia	444.8
	Japan	37.7	Malaysia	78.0	United States	107.2
Coal, coke, and briquettes (D32)	North Korea	58.9	North Korea	48.5	North Korea	27.2
	Australia	12.0	Australia	22.4	Australia	7.8
	Hong Kong	Negl.	Canada	5.6	Canada	4.7
Gas, natural and manufactured (D34)	Hong Kong	1.2	Hong Kong	4.0	U. Arab Emirates	18.1
	Singapore	1.2	Singapore	2.9	Singapore	12.4
	United States	0.7	Indonesia	2.4	Japan	11.7

Notes: Negl. negligible. Figures are rounded.

[a] The Soviet Union (USSR) formally dissolved on 31 Dec. 1991, paving the way for the formation of a new Commonwealth of Independent States (CIS).

Sources: General Administration of Customs, PRC (1983–92); United Nations (1984–90); Minerals Policy Program (1993).

TABLE 5.8. *China's mineral trading partners, 1987–1991*

	1987		1989		1991	
	Major countries	Values (US$m.)	Major countries	Values (US$m.)	Major countries	Values (US$m.)
EXPORTS						
Crude fertilizers and minerals (D27)	Japan	144.3	Japan	232.7	Japan	193.9
	Hong Kong	27.4	Hong Kong	63.5	Hong Kong	65.2
	United States	50.1	United States	57.1	EC	60.3
Metalliferous ores and scrap metal (D28)	Soviet Union	40.9	Japan	67.8	Japan	33.8
	Singapore	39.9	Hong Kong	61.3	EC	31.3
	Japan	31.1	United States	54.8	United States	28.4
Non-metallic mineral manufactures (D66)	Hong Kong	189.5	Hong Kong	357.8	Hong Kong	605.2
	Japan	40.0	Japan	80.8	EC	13.5
	United States	36.5	United States	56.2	Japan	13.1
Iron and steel (D67)	Japan	125.6	Japan	174.3	Japan	414.2
	Hong Kong	105.0	Hong Kong	153.3	Hong Kong	324.6
	Holland	35.2	Thailand	78.0	Thailand	173.9
Non-ferrous metals (D68)	Hong Kong	179.2	Hong Kong	150.3	Hong Kong	481.4
	Japan	153.6	Japan	117.4	EC	299.4
	United States	108.6	United States	78.6	United States	249.2
Manufactures of metals (D69)	Hong Kong	186.2	Hong Kong	300.1	Hong Kong	481.4
	FRG[a]	36.7	United States	201.7	EC	299.4
	Singapore	35.2	Singapore	52.4	United States	249.2

IMPORTS						
Crude fertilizers and minerals (D27)	Morocco	12.0	Morocco	11.8	Morocco	13.6
	United States	1.6	Canada	10.6	Soviet Union	6.9
	Soviet Union	1.3	United States	8.8	Hong Kong	5.3
Metalliferous ores and scrap metal (D28)	Australia	229.7	Australia	348.1	Australia	609.2
	United States	47.0	Brazil	74.4	Brazil	101.5
	Canada	32.8	India	67.8	United States	96.8
Non-metallic mineral manufactures (D66)	Japan	84.4	Japan	191.5	Japan	179.3
	Hong Kong	67.4	Hong Kong	129.8	Hong Kong	100.2
	North Korea	45.5	Burma	29.7	EC	53.1
Iron and steel (D67)	Japan	2,520.8	Japan	2,750.4	Japan	1,588.1
	FRG	415.1	FRG	520.4	EC	319.8
	Soviet Union	367.9	Soviet Union	398.1	FRG	185.5
Non-ferrous metals (D68)	Soviet Union	165.5	Soviet Union	216.0	EC	140.8
	Japan	106.8	Japan	119.0	Hong Kong	122.1
	Australia	58.2	United States	90.6	Japan	107.6
Manufactures of metals (D69)	Japan	174.6	Hong Kong	192.5	Hong Kong	191.3
	Hong Kong	131.9	Japan	119.6	Japan	98.3
	United States	30.6	France	45.2	EC	55.1

Note: Figures are rounded.

[a] West Germany, or the Federal Republic of Germany, and East Germany, or the German Democratic Republic, officially reunited on 3 Oct. 1990, forming a new Federal Republic of Germany (FRG).

Sources: General Administration of Customs, PRC (1983–92); United Nations (1984–90); Minerals Policy Program (1993).

majority of Chinese mineral exports in 1991, namely Hong Kong (ranked first in value terms), Japan (second), the EC (third), the United States (fourth), and Thailand (fifth). Although Hong Kong is the major recipient of Chinese mineral goods, much of its imports are further processed, and subsequently re-exported to their final destinations. As the leading importer of Chinese non-metallic mineral manufactures, non-ferrous metals, and manufactures of metals, in 1991 Hong Kong imported nearly $1.32bn. worth of these commodities, much of which were subsequently re-exported. Japan, China's second largest overall trade partner, received more crude fertilizers and minerals and iron and steel from China than all other nations of the world. Being a resource-poor nation, Japan depends on mineral imports from countries like China to meet the raw material requirements of a major industrial nation.

China imports substantial minerals from throughout the Asia–Pacific region as well as the EC, France, Germany, Brazil, and Morocco. The most valuable imports are of iron and steel, emanating principally from Japan ($1,588.1m.), the EC ($319.8m.), and Germany ($185.5m.). An important structural change in China's mineral trade flows during the 1980s was the dramatic increase in value of iron and steel imports from Japan and other nations. In 1980, for example, China imported just $1.6bn. worth of iron and steel from Japan, whereas by 1989 these imports grew substantially in value to $2.75bn. to fall the following year. The dramatic increase through much of the decade was needed by China to meet its over-ambitious industrialization programme as set forth in the Sixth and Seventh Five-Year Plans. Generally speaking, China depends heavily on Japan for its modernization programme. Owing to its geographic proximity, Japan is better suited to trade with China than other Western nations like the United States and Germany. Lower transportation costs of capital goods and maintenance expenses of Japanese-produced machinery make Japanese products more attractive to China.

A growing mineral import source for China within the Asia–Pacific region is Australia, with its vast resources of iron ore, refined lead and zinc, and aluminium. China is increasingly becoming a major market for Australia's mineral resource export sector, and is presently the sixth largest market for these commodities (Bureau of Agricultural and Resource Economics 1988). If China's import requirements continue to expand, Australia will be in a position to increase its exports of iron ore, aluminium, crude steel, and refined zinc. However, because other nations in the Asia–Pacific region are also interested in expanding their export supplies to China, Australia's competition for exports to China will probably remain strong.

To visually compare and contrast the major export and import markets of China, Figure 5.7 shows the top ten destinations and sources of Chinese mineral and energy exports and imports in 1991. The two diagrams include values of all mineral and energy commodities. Japan is clearly the major recipient of Chinese mineral and energy exports, particularly of petroleum and petroleum products,

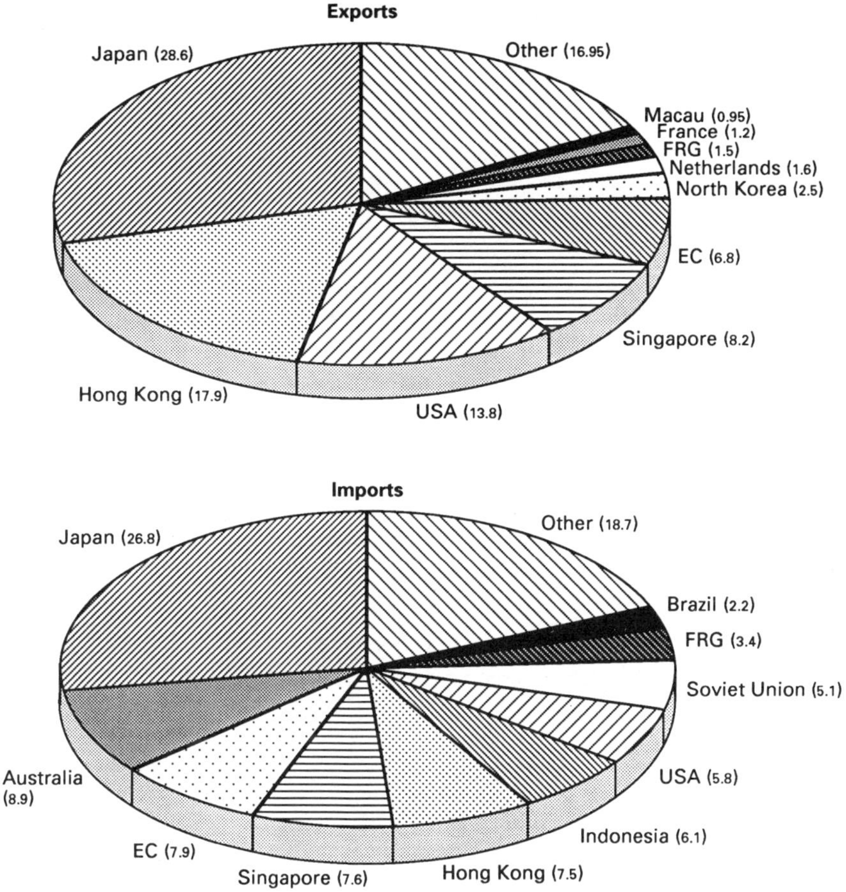

FIG. 5.7. China's major mineral and energy trading partners, 1991

coal, coke, and briquettes, non-ferrous metals, and crude fertilizers and minerals. Other countries receiving substantial mineral and energy exports from China include Hong Kong (17.9 per cent), the United States (13.8 per cent), and Singapore (8.2 per cent).

In terms of import sources, in 1991 Japan accounted for 26.8 per cent of the total value of imported goods received by China. This included $1.59bn. worth of iron and steel exports. Australia, the EC, Singapore, and Hong Kong also accounted for a significant percentage of China's imports that same year.

If the value of China's petroleum exports is removed from the pie diagram, the top ten country destinations change noticeably, reflecting the extraordinarily large value of petroleum exports to Japan. Likewise, if the value of iron and steel imports is removed from the import diagram, the sources change as well. Figure 5.8

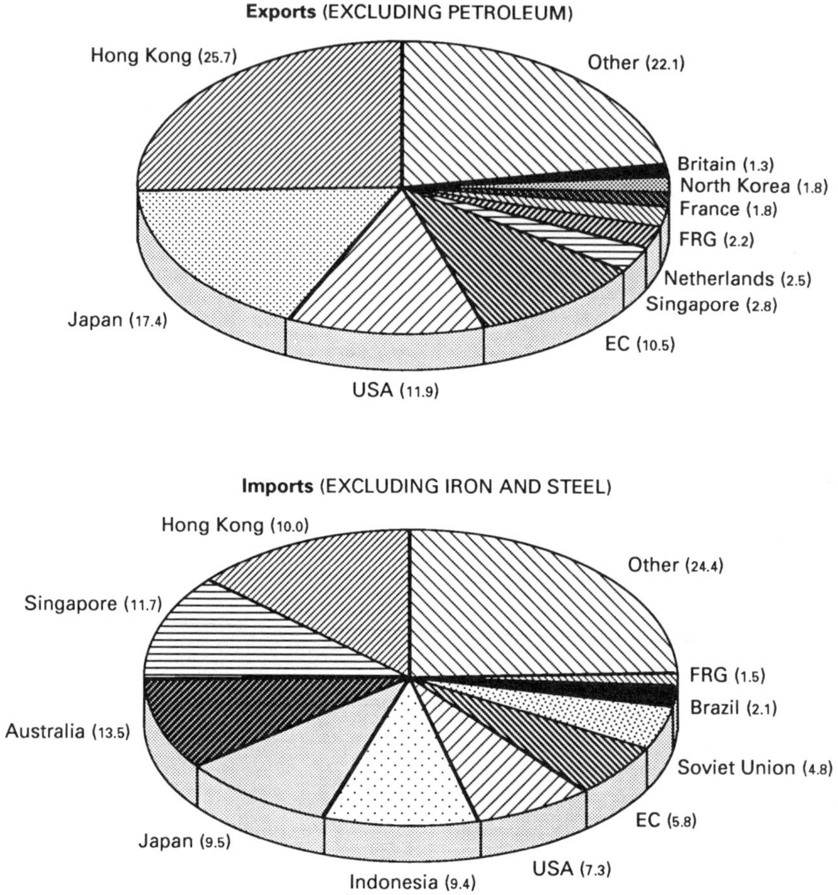

FIG. 5.8. China's major mineral and energy trading partners (exclusive of selected commodities), 1991

illustrates these changes. Of particular note is that in terms of imports, Japan's value contribution drops from 26.8 to 9.5 per cent when the value of iron and steel is removed, while the relative shares of import value of those countries which export products other than iron and steel to China rise.

MINERALS AND ENERGY TRADE IN THE 1990S

China's foreign trade system reform began in 1979 primarily by the decentralization of foreign trade management. Centralized controls over trade activities, previously supervised by the Ministry of Foreign Economic Relations and Trade

(MOFERT), were relaxed in 1984, only to be reimposed after a jump in imports the following year (World Bank 1990*b*). During most of the 1980s enterprises and departments of industrial production and local governments at a variety of levels were permitted to participate in export/import practices instead of the centralized management by national foreign trade corporations under MOFERT. During the 1990s, under the guidance of China's Eighth Five-Year Plan and Ten-Year Programme for economic development, the nation will continue to expand efforts at promoting exports and shifting from selling lightly processed goods to exporting intensively finished products that are competitive internationally. Official policies on imports will continue to stress limits on consumer goods and simple manufactures, but provide increased attention to the need to boost imports of high technology and materials for export-oriented industries. Various trade measures adopted during the 1980s and 1990s have already led to substantial changes.

While reform of China's foreign trade regime led to increased trade activities between China and the international community in the 1980s, the country's foreign trade deficit worsened steadily in the early part of the decade, dropping to a low of $14.9bn. by 1985. A gradual improvement in the trade balance occurred in the latter part of the 1980s, with the deficit shrinking to $6.6bn. in 1989. In 1990 China's trade performance reversed, with the nation registering a significant current account surplus for the first time since 1983. Importantly, export growth provided much of the stimulus to China's economy in 1990. China's growing number of enterprises financed by overseas investment were well positioned to take advantage of their access to export markets by the end of the 1980s, stimulated in large part by favourable exchange rate policies. China's trade surpluses in 1990 and 1991 reached $87.1bn. and 81.2bn. respectively, contributed largely by tremendous export growth.

Though not necessarily specific to mineral or energy enterprises, many of the reforms instituted within China's foreign trade system during the many years of decentralization and reform influenced the trade abilities and prospects of China's many trading organizations. Key reforms included expansion of the contract-responsibility system for foreign trade, the granting of permission for trading organizations to retain a higher proportion of foreign exchange earnings, greater tax remissions for exports, and greater autonomy for foreign trade organizations. The years immediately ahead provide a mixed climate for any additional loosening of controls on China's trade enterprises, as the recent strong economic growth in 1991 and again in 1992 have led to worries among China's leaders that the economy is on the verge of overheating, forcing China once again to confront the spectre of runaway inflation. Continued anxiety over growing imports and exports may lead to further tightening of import controls on consumer goods and raw materials, in an attempt to curtail the outflow of precious foreign exchange, and a curtailing of Chinese enterprises now authorized to trade directly with foreign companies. In 1989–90, an estimated 1,300 foreign trade corporations were disbanded in China as government authorities sought to recentralize some control over export/import companies in an attempt to curb escalating trade practices.

In evaluating likely characteristics of China's long-term minerals or energy trade, the possible restoration of China's membership in the General Agreement on Tariffs and Trade (GATT) must also be considered. GATT, a 97-nation compact which administers global trading rules, is the world's leading trading organization. With China becoming more developed in its international trade and economic relations, in July 1986 the government requested re-entry into GATT. Progress on restoration of membership has been and will continue to be slow, although negotiations between Chinese and GATT officials are continuing. When China originally applied for GATT membership the United States and the European Community supported the decision; however, the reintroduction of central controls on the economy in late 1989 set the application back. Today, China plans to further reform its foreign trade administration as it moves closer to rejoining GATT. China expects to reduce the number of export products now under rigid control substantially in the future, and work on regulations that will drastically cut both tariff and non-tariff barriers. The redefining of China's rights and obligations as a contracting member of the 97-nation international trade body is the last major work to be done before China acquires GATT membership.

With minerals and energy trade being so important to China, GATT officials have posed many questions to the Chinese government about proposed changes in the trade practices of the mineral and energy industries. Other concerns to GATT officials include government subsidization of the industries, state ownership of resources, planned production targets, and pricing policies. China's re-entry into GATT will be contingent upon many changes in trade and other economic practices according to the provisions of GATT. It is difficult to predict how long it will take GATT officials to re-establish China's membership, but it will involve a great deal of bilateral negotiation between China and many member countries on tariff concessions so as to help boost trade on each side.

China's energy trade surplus in 1991 was $2.9bn., primarily resulting from the high value of petroleum and petroleum export earnings. Although China's petroleum exports constitute only a small percentage of total world oil transactions, the changes in export quantities and values since 1961 (the first year of refined petroleum product exports) have had a strong impact on China's economy. Rising oil export revenues until 1986 indeed helped to finance China's ambitious modernization programme during the past decade. In 1986, when worldwide oil prices plummeted and cut China's oil export earnings nearly in half, the Chinese government was dramatically awakened into understanding the vulnerabilities of over-reliance on a single commodity. By good fortune, in 1986 and again in 1987, China's textile export earnings rose dramatically, helping to reduce the negative aspects of the decline in oil export revenues at that time.

The future of petroleum exports by China will be dependent on a host of factors including growing domestic demand, continued needs for foreign exchange, changes in production levels, and counter-trade agreements. With regard to the latter, counter-trade deals in which oil is used as payment for imported technology or equipment are likely to remain an important component of future foreign trade. In the light of growing domestic demand for oil and a slow-down in production growth, by the mid-1990s the volume and direction of petroleum trade will change significantly. Table 5.9 clearly illustrates China's growing reliance on imported crude oil and products, which has increased nearly fivefold since 1982. China's crude production is likely to remain relatively stagnant until 1995, after which output will depend on whether the Daqing field starts declining and whether new fields are discovered and developed.

According to Fridley and Fesharaki (1987*b*), 1985 marked a peak in petroleum product export income, and thus was a turning point in China's dependence on oil export earnings. Fridley (1988; 1991*d*) also estimated that China will become a significant importer of petroleum products by the mid-1990s, assuming that Chinese government officials allow imports to satisfy domestic needs. The extent of these imports will depend on the speed at which China's refining industry will expand.[1] Based on a variety of assumptions on output, refinery expansion, and demand, Table 5.10 presents a forecast for China's oil supply and demand balance to 2000, highlighting the implications for trade (Fridley 1991*d*). Though the high production/low demand scenario for 2000 appears to be in balance, the critical factor in relating supply to demand is refining. In even the most optimistic projections whereby China constructs a new 100,000 b/d refinery every year until 2000, a shortfall in refinery capacity will limit product output to around 3m. b/d (Fridley 1991*d*).

China's coal exports rose from 5m. tonnes in 1980 to 20m. tonnes in 1991, and are projected to increase to about 30m. tonnes in 2000 and 40m. tonnes by 2010.[2] Although the potential exists for increased Chinese coal exports to the Asia–Pacific region, there will be only gradual gains in exports to this as well as other regions over a period of years. The nation's dependence on coal as its major energy source is the overriding reason for this. China is presently the third largest energy consumer in the world, with much of this being met by domestic coal resources. Growing electricity needs in the country also suggest increased usage of coal, as coal-fired power stations are the predominant source. Additional constraints include the continued technical and financial problems at China's northern Antaibao mine, where most coal produced is intended for export. US-based Occidental Petroleum Corporation, once a major partner in the huge Antaibao opencast mine, baled out in 1991 just as the mine was beginning to produce results after almost eight years of frustrating development. Occidental's withdrawal has helped to discourage further foreign investments in similar large coal projects. It is unclear whether the lack of interest by Western companies in Chinese coal-mine developments will hamper the country's efforts to boost exports.

TABLE 5.9. *China's oil trade balance, 1982–1991*

	Crude production	Exports			Ratio exports/ production	Imports		
		Crude	Products	Total		Crude	Products	Total
1982	102.10	14.68	4.91	19.59	0.19	0.64	0.93	1.57
1983	106.10	14.83	4.91	19.74	0.19	0.37	0.98	1.35
1984	114.5	22.01	5.70	27.71	0.24	0.25	0.88	1.13
1985	124.90	30.03	6.21	36.24	0.29	0.72	0.90	1.62
1986	130.69	28.5	5.46	33.96	0.26	1.72	3.23	4.95
1987	134.14	27.23	4.94	32.17	0.24	1.72	3.23	4.95
1988	137.05	26.05	4.79	30.84	0.23	1.96	4.22	6.18
1989	137.64	24.39	4.74	29.13	0.21	3.26	5.34	8.6
1990	138.30	23.99	5.26	29.25	0.21	2.83	2.7	5.53
1991	139.60	22.6	4.81	27.41	0.20	5.97	4.61	10.58

Sources: State Statistical Bureau (1981–92: various issues); General Administration of Customs, PRC (1987–92: various issues); State Statistical Bureau (1982–8: various issues); British Petroleum (1984–92 (June 1992)).

TABLE 5.10. *Outlook for petroleum supply, demand, and trade*

	1980	1989	1990	1991		1995		2000	
				Low	High	Low	High	Low	High
Crude supply									
Onshore									
Daqing	1,030	1,110	NA	1,095	1,100	900	1,000	NA	NA
Shengli	352	667	NA	680	700	800	900	NA	NA
Others	737	956	NA	982	955	1,111	991	NA	NA
Offshore	0	19	24	30	36	60	100	75	140
TOTAL	2,119	2,752	2,766	2,787	2,791	2,870	2,990	3,150	3,510
Demand	1,472	2,053	2,156	2,242	2,285	2,830	3,110	3,540	4,480
Refinery output	1,556	2,041	2,198	2,242	2,275	2,450	2,590	2,800	3,320
Imports									
Crude	0	64	58	64	90	0	200	0	560
Products	0	107	63	90	110	240	710	190	1,640
Exports									
Crude	266	488	480	440	468	0	270	0	270
Products	84	95	105	90	110	0	50	0	0

Source: Fridley (1991*b*).

As for minerals, in a widely expected move Chinese government officials in early 1989 banned the export of six 'scarce' metals and limited the export of 12 others (*Mining Journal*, 23/30 Dec. 1988). The bans and restrictions went into effect on 1 Jan. 1989; rigid central controls on non-ferrous metals trading were eased somewhat by 1991.

Until the ban, China appeared to be willing to export selected mineral (and energy) commodities to earn foreign exchange, even at the expense of not meeting the needs of domestic industries. But the dramatic clamp-down suggested a flexible policy where given certain conditions efforts would be made to provide domestic industries with their needed raw materials, curtail the spending of hard currency on mineral imports, and capture the value added by processing. The recentralization of trade and restriction of selected imports in 1990 also illustrated the willingness of Chinese government officials to impose additional controls on imports of raw materials.

Non-ferrous metals included in the 1989 export ban were copper, aluminium and nickel and their alloys, and platinum, while only limited exports of chromium and molybdenum ore, magnesium and manganese metal, ferrochrome and ferromanganese were allowed. The clamp-down on exports had been expected for some time owing to the increased demand for these materials by China's own domestic industries, and the severe shortages of foreign exchange in the country. Imports of the scarce materials were restricted as well because of the general shortage of foreign exchange in China by the end of the 1980s.

The extent of disruption to China's overall trade pattern resulting from the 1989 ban and temporary slow-down of trade reform was shown in Tables 5.2 and 5.3. Exports of non-ferrous metals from China were nearly halved in 1989, amounting to US$453.7m. Imports of non-ferrous metals did not respond to controls until 1990, when their value dropped by more than $535m. from the preceding year. While the impacts of the 1989 trade measures are still being felt in China, the nation's rapid industrialization and economic growth, particularly in southern China, have defied most attempts to place restrictions on metals trade. In 1993 the Chinese government implemented another partial ban on the export of selected metals—copper and its alloys and platinum—with the hope of slowing trade of those commodities which are in short supply.

Of all non-ferrous metals, China's growing demand for copper is most troublesome to industry leaders, as imports continue to rise almost unabated. It is the profits from copper-trading that has attracted larger and larger numbers of semi-private or private enterprises in China to trade metals (*Far Eastern Economic Review*, 16 Apr. 1992). China's net imports of copper grew from 70,082 tonnes in 1989 to 166,134 tonnes in 1991. Copper imports will continue to increase in the next few years. Indeed, by the end of the century, imports are expected to reach 300,000 tonnes, compared with the near 200,000 tonnes projected for 1992 (*China Daily*, 18 Jan. 1993).

Much of China's imports of copper as well as other non-ferrous metals are through Hong Kong, which acts as a principal conduit for this demand. Because of their growing wealth and stature, industrial companies in southern China are increasingly purchasing their own raw materials on the international market through Hong Kong. Officials in Beijing have attempted to discourage copper imports by imposing a 23 per cent tariff on imports, while continuing duties on exports on zinc, for which the nation has a surplus. Chinese tariff rates are traditionally dependent on whether in the government view China has a surplus or a shortage of a particular commodity.

In an effort to reduce mineral imports and perhaps increase exports over the long term, China will continue boosting output of its major mineral commodities. As evidence of this, the China National Nonferrous Metals Industry Corporation (CNNC) recently projected that its output of non-ferrous metals in China would rise from 2.35m. tonnes in 1990 to 3.5m. tonnes in 1995, and 5.0m. tonnes by the end of the century (*China Daily*, 18 Jan. 1993). This suggests an increase in output of these metals of more than 200 per cent between 1990 and 2000. The major non-ferrous metals referred to in these projections include copper, aluminium, lead, zinc, nickel, tin, antimony, mercury, magnesium, and titanium.

China's tremendous demand for raw materials has meant Chinese buyers are becoming increasingly important players in many of the world's non-ferrous metals markets. China's metal traders are gradually joining forces with large domestic production firms to set up international export bases. Companies, such as the China National Nonferrous Metals Import and Export Corporation (CNIEC), are also relying on overseas affiliates to expand business worldwide. While relying on its traditional footholds in the Asia–Pacific region, Europe, and North America, CNIEC is exploring opportunities for business with the Middle East, Latin America, the former Soviet Union, and Eastern Europe. The company will not limit itself to only exporting Chinese products or importing foreign products for Chinese firms, but it will also begin acting as an international trader. CNIEC as well as other Chinese trading firms are now considering purchasing one country's products for resale to another country. They will continue border and barter trade, while also becoming involved in futures trading.

In many parts of the world one of the greatest influences on non-metallic minerals trade in recent years has been the export of a variety of goods to international markets from China. The nation now exports 55 kinds of non-metallic minerals, including talcum, fluorite, magnesite, graphite, and construction stone. Although the value of non-metallic mineral exports is only $300m. at present, this figure has doubled in five years. Prospects for increased exports of non-metallic minerals are clearly favourable.

Under the guidelines of China's recently implemented Eighth Five-Year Plan (1991–5), several new cement plants are scheduled for renovation or construction during the period, while additional fertilizer plants are also slated for construction. Though China is the world's largest producer of cement, it had been a major importer of the commodity during the 1980s until efforts to boost output proved successful in reducing the need for foreign supplies. China's output of cement is still insufficient to meet the demands of a growing industrial base, and has therefore slowed or halted the initiation of many construction projects. China relies heavily on foreign nations for technology and equipment for the making of cement, as well as glass and other building materials.

MAJOR PATTERNS

China's minerals and energy trade is an important component to economic growth and expansion, and plays a major political role in helping to improve China's relations with neighbouring countries, as China produces many metals which are required by them. With regard to metals, due to a limited resource-processing industry, China has adopted a 'two-ends out' (*liang duan zai wai*) policy of importing raw materials and exporting much of the product (Clark, Dorian, and Fridley 1989). Alumina, for example, is imported from Japan and processed into aluminium for export back to Japan, with retention of some of the import or export as partial payment for the processing. Identical activities are under way in the copper and steel industries.

Unlike the metal and non-metal industries where China's low labour costs provide comparative advantage, the energy-processing industries are highly capital-intensive. China, therefore, is less likely to develop into a regional refining centre, although there have been a number of proposals to build export refineries on China's south-eastern coast. The high cost of new refineries, in combination with low international oil prices, makes the economics of such projects highly questionable.

China's overall minerals and energy trade balance in 1991 was positive owing to the greater than $2.8bn. surplus revenues earned through energy trade. China's petroleum export revenues have traditionally represented a major portion of overall export earnings (around one-fifth to one-quarter), although such revenues have declined since 1985. China's energy trade balance has been positive since 1970 and reached a peak in 1985 when petroleum and petroleum product export earnings were an estimated $6.63bn. It was during the late 1970s and early 1980s when China's oil export earnings grew substantially as a result of increased production, better refining capabilities, and strongly boosted world oil prices.

In contrast to energy trade, China's minerals trade deficit steadily increased from 1970 to 1986 at a strong rate. After 1986, the trade scenario improved gradually until reaching a positive balance in 1991. In 1970 the deficit ran an estimated $450m.; by 1986 the figure had reached $7.61bn. Much of the trade deficit during these years was due to the enormous value of imported iron and steel into China, which peaked at $6.65bn. in 1986. An improvement in the trade scenario towards the end of the 1980s required substantially decreased imports of iron and steel. Therefore the Seventh Five-Year Plan of China facilitated a massive renovation of China's iron and steel industry to boost domestic capacity and help shrink imports.

The various tables and figures presented in this chapter reveal much about recent trade patterns of China. These include:

1. China's minerals and energy trade in 1991 was valued at $29.86bn., or nearly 22 per cent of the overall trade value.
2. China's minerals and energy trade is largely restricted to Pacific Basin nations, particularly Japan, Hong Kong, the United States, Singapore, and Australia.
3. Nearly 18.5 per cent (value terms) of China's metals and minerals trade and 4.8 per cent of the nation's energy trade flows through Hong Kong, where much of the materials are further processed and/or re-exported. Hong Kong is generally considered a trans-shipment port for Chinese shipments.
4. Export earnings of petroleum and its products reached a peak (in value) in 1985.
5. Although some potential exists, the prospects for substantial increased coal exports from China are slim.
6. China's minerals trade balance steadily improved from 1985 reaching a surplus in 1991, due largely to a rapidly rising value of non-metallic mineral manufacture exports.
7. The growing deficit in metalliferous ores and scrap metal may jeopardize China's positive minerals trade balance.
8. Exports of non-metallic mineral manufactures, iron and steel, non-ferrous metals, and manufactures of metals are on the rise.
9. Imports of four of eight mineral groups rose between 1989 and 1991, but not enough to offset the dramatic declines in imports of iron and steel and non-ferrous metals.
10. China's limited resource-processing capabilities necessitate the exporting of some ores and concentrates and the importing of some semi-finished products.

Given the proposed increases in minerals production, Chinese government officials are confident that the nation's minerals trade balance will remain positive in the years ahead. Much of this will be determined by changes in trade flows of petroleum and petroleum products and iron and steel. A conflict may emerge, however, between China's ambitious economic goals and its domestic resource requirements of metals and energy. Problems may arise as the need to export minerals and energy to earn foreign exchange will conflict with the country's domestic resource requirements to sustain economic growth. Simply stated, China cannot be considered a resource-rich country with nearly unlimited mineral and energy resources when the per capita reserve amounts are examined as well as the resource needs to satisfy a growing industrial demand. China is therefore now in the process of adjusting from an economy whose resources were perceived by many to be rich, to one in which selected resources are indeed in short supply owing to a number of constraints.

OTHER FORMS OF INTERNATIONAL CO-OPERATION

1. A Pronounced Need

Minerals and energy trade is one of many means of linking the Chinese economy to the international community. With the advent of the open-door policies in 1978, China's level of co-operation with neighbouring Asia–Pacific economies as well as countries beyond the Pacific Rim began to grow. Such co-operation has included, beyond commodity trade, foreign borrowing and financing, science and technology transfers, joint-venture activities, and a host of other forms of economic linkage. China's emergence into the global economy since initiating the policy of opening up to the outside world has been dramatic and robust. Despite this, however, China's international co-operation is still at an infant stage.

Today, China's foreign trade represents only a small percentage of intraregional trade, while total foreign direct investment accounts for an equally small percentage of Japanese and US investment in the region. Continuing efforts by the Chinese government to expand co-operation with the international community, such as joining economic organizations like the Asian Development Bank and pursuing membership in GATT, suggest that China's participation in and interaction with the world community will grow. Chinese government officials realize the importance of expanding trade and encouraging the influx of foreign capital, technologies, and skills in order to achieve its ambitious development objectives. China's leaders have repeatedly assured potential foreign investors that any efforts to stabilize its domestic economy in the years ahead will not curb the country's ambitions to expand foreign economic relations and trade, nor will the implementation of joint investment and co-operative venture agreements or foreign government loan programmes be disrupted.

2. Foreign Investment, Involvement

Utilization of foreign capital is a critical component of China's long-term strategic policy for socialist economic development. With the conclusion of the third plenary session of the Chinese Communist Party's Central Committee in late December 1978, the introduction of capital from abroad has represented one of China's main avenues of opening up to the outside world. The 12th National Congress of the CCP in September 1982, subsequent fifth session of the Fifth National People's Congress in late 1982, and the issue of the Provisions on the Encouragement of Foreign Investment in October 1986 further established and strengthened foreign investment as a primary means of linkage to other economies.

China employs two methods to bring foreign capital into the country, either by borrowing money from foreign countries or attracting direct foreign investment

(DFI).[3] Attracting DFI includes the creation of joint ventures, co-operative business operations, co-operative development, enterprises owned exclusively by foreign capital, compensation trade, the processing of materials supplied by clients, and the assembling of parts and components supplied by them. In the statistical reporting on DFI, Chinese government authorities categorize investment as one of four types—the equity joint venture, contractual joint venture, wholly foreign-owned joint venture, and joint oil development projects. An equity joint venture is a limited liability company jointly managed by the investors in proportion to their respective shares of investment. Equity joint ventures are preferred over other types of DFI by the Chinese government, and now represent the predominant vehicle for DFI in the country.

During the 1980s, China's use of foreign capital increased yearly. The trend continued into the 1990s. By the end of 1991, China's total foreign loans had reached US$52.5bn., including foreign commercial loans (*Beijing Review*, 10–16 Aug. 1992). Newly registered foreign investment set new record highs again in 1992 in terms of both the number of overseas-funded projects and pledged capital inflow. By 1993, more than 40,000 foreign-funded enterprises had been approved in China, with three-quarters having been put into operation. Foreign capital sources have grown more diverse in recent years, and the Chinese government continues to encourage foreign investment to fund development projects.

Table 5.11 presents direct foreign investment figures for China during the years 1979–91. The table disaggregates values by equity joint ventures, co-operative ventures, wholly foreign-owned ventures, and joint oil development projects, four of six categories of DFI recognized by the Chinese government, the other two being compensation trade and imported inputs for processing and assembly. All figures are in US dollars. Figure 5.9 graphically depicts the data shown in Table 5.11.

Long-term DFI values clearly reveal that China has achieved a steady and expanding development in the utilization of foreign capital since 1979. By 1985, pledged foreign investment reached $4.6bn., nearly three times the value offered at the beginning of the decade. In the early to mid-1980s, the amount of foreign capital pledged was much larger than actual foreign investment owing to uncertainty surrounding legislation on joint ventures and numerous bureaucratic complications. While direct foreign investment fell dramatically in 1986, it rebounded in 1987 and 1988 to exceed its 1985 level.

By the end of the 1980s, China was enjoying a boom in foreign investment to coincide with strong economic growth. From January 1987 till June 1989—a period of civil unrest and turmoil—the number of signed contracts increased 400 per cent, and contract value grew 67 per cent over the totals accumulated from 1979 to 1986 (Brecker 1990). After the initial shock of June 1989, foreign investment picked up once again. By year-end 1989, $5.6bn. of foreign capital was pledged, more than 5 per cent greater than the preceding year. A majority of this investment was, however, committed in the first half of the year.

TABLE 5.11. *Direct foreign investment in China, 1980–1991*

	1980	1981	1982	1983	1984	1985	1986	1987	1988	1989	1990	1991
Equity joint ventures												
Ventures approved (no.)	20	28	29	107	741	1,412	892	1,399	3,909	3,659	4,091	8,395
Foreign investment pledged[a]	63	28	29	188	1,060	2,030	1,375	1,920	3,134	2,659	2,704	6,080
Actual foreign investment[a]	NA	NA	100[b]	74	255	580	804	1,480	1,975	2,037	1,836	NA
Co-operative ventures												
Ventures approved (no.)	320	70	402	331	1,089	1,500	582	786	1,621	1,179	1,317	1,778
Foreign investment pledged[a]	500	1,300	926	504	1,480	2,189	1,358	1,286	1,624	1,083	1,254	2,138
Actual foreign investment[a]	NA	NA	531[b]	227	460	580	790	620	780	742	674	NA
Wholly foreign-owned ventures												
Ventures approved (no.)	4	14	12	18	26	46	18	45	410	931	1,861	2,795
Foreign investment pledged[a]	14	262	55	44	79	32	20	470	481	1,654	2,444	3,667
Actual foreign investment[a]	NA	NA	40[b]	43	15	13	16	25	226	371	683	NA
Joint oil development projects												
Ventures approved (no.)	4	None	1	18	None	4	6	3	5	10	5	10
Foreign investment pledged[a]	1,112	None	170	1,031	None	360	80	4	59	204	194	92
Actual foreign investment[a]	NA	None	497	292	520	480	260	180	212	232	244	NA
TOTAL												
Ventures approved (no.)	348	112	444	474	1,856	2,962	1,498	2,233	5,945	5,779	7,274	12,978
Foreign investment pledged[a]	1,689	1,590	1,180	1,767	2,619	4,611	2,833	3,680	5,298	5,600	6,596	11,977
Actual foreign investment[a]	NA	NA	1,168[b]	636	1,250	1,653	1,870	2,305	3,193	3,392	3,437	4,370[c]

Note: NA not available.

[a] In $US million.

[b] Total for 1978–82, $US million.

[c] Estimated based on 1990 percentage shares of actual foreign investment.

Sources: State Statistical Bureau (1981–92); Pomfret (1991, 1989); Lee *et al*. (1990); MOFERT (1986–91: various issues); Minerals Policy Program data files.

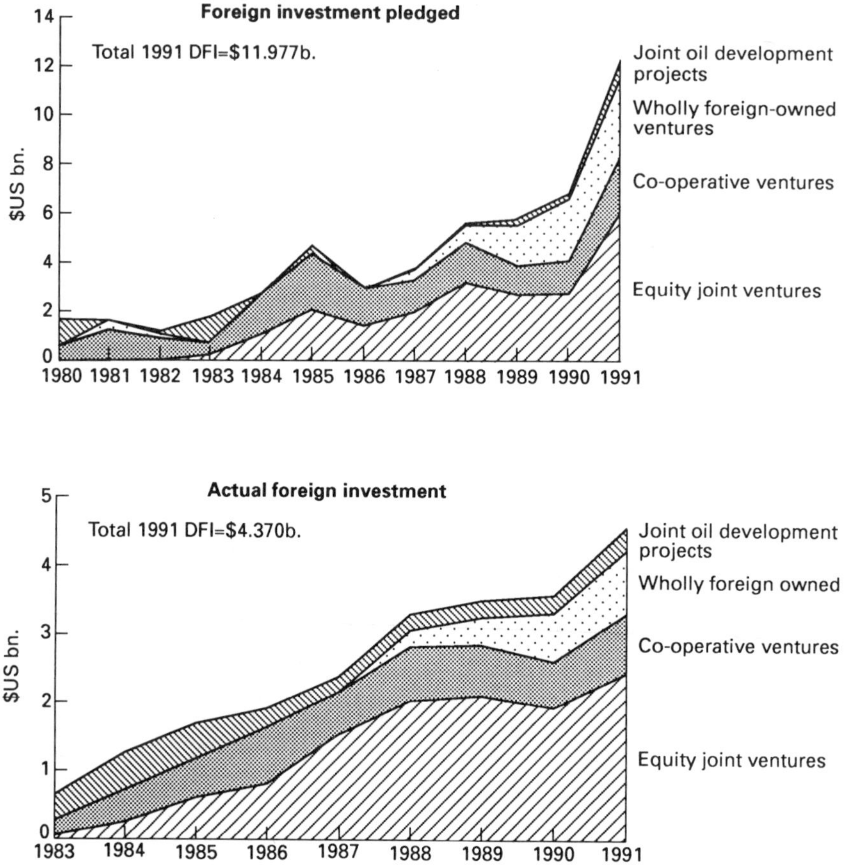

FIG. 5.9. Direct foreign investment in China, 1980–1991

Though China's investment boom was temporarily broken in 1989, foreign investment in China set new records in 1990, more than one year after the Beijing student/military confrontation. Official figures reveal that 7,274 ventures were approved, with a total of nearly $6.6bn. of investment pledged. The increase in pledged investment suggests that investors continue to have a relatively favourable view of overall investment conditions in China, despite the domestic turmoil and three-year economic austerity programme. Actual foreign investment in 1990—$3.4bn.—slightly outpaced the 1989 level. The biggest increase in investment was in the wholly foreign-owned ventures category, where the number of projects approved doubled from 1989, and actual foreign investment rose by more than 45 per cent.

Foreign investment continues to reach China today at record levels. On a contract basis, direct investment projects in the country numbered 12,978 in 1991 (up 78.4 per cent from a year earlier) and were valued at $11.98bn. (up 81.6 per cent). On a realization basis, direct investment totalled $4.37bn. (up 25.2 per cent). In both cases, the figures represented record highs. Between 1990 and 1991 the foreign investment pledged for equity joint ventures rose by 125 per cent, for co-operative ventures 70 per cent, and for wholly foreign-owned ventures 50 per cent. Foreign capital pledged for joint oil development projects dropped by just more than $100m.

Foreign investment into China also surged in 1992, with newly registered investment setting new record highs in terms of both the number of overseas-funded projects and pledged capital inflow. Actual foreign investment reached $8.34bn., or almost twice as much as in the whole of 1991, or more than one-quarter of the funds invested in China since 1979, the first year in which Deng Xiaoping's open-door policies were implemented. The total number of ventures approved skyrocketed to 40,291. China is seeking a target of a cumulated total of $40bn. in DFI by the end of this century by attracting at least $25bn. during the 1990s (Renmin Ribao 1990).

While the growth of direct foreign investment in China has been tremendous in many ways, neither the amount of investment nor the transfer of advanced technologies have met the original objectives. Investors still do not enjoy complete access to China's local market, and they continue to struggle with bureaucratic complications, foreign exchange and finance difficulties, supply problems, high fees and charges, and local interference. To alleviate some of these problems, in April 1990 the Chinese government approved the Amendment to the Law on Joint Ventures Using Chinese and Foreign Investment. The change allowed foreign partners to exercise a more important role in management and protected-equity joint ventures against expropriation by the state. China has also begun leasing land on a long-term basis (up to 70 years) in special economic zones and open cities. Given a gradual improvement in relations with many industrialized countries and continued robust economic growth, a rise in DFI in the years ahead can be expected.

Despite there being nearly 40,300 foreign-funded business ventures set up in China since 1979, relatively few of these have been within the mineral and energy industries. Among these two industries, by far the majority of foreign investment has gone into the energy industry. In encouraging the establishment of joint ventures with foreign capital, China gives energy development the highest priority. At present, most foreign capital in this sector is used in offshore oil prospecting and coal exploitation.

Foreign companies have invested more than $4.2bn. in the search for offshore oil in China since the area was first opened to Western companies in 1979. During that year, China had concluded agreements with more than 15 oil companies from the United States, the United Kingdom, France, Japan, and Italy to search for oil reserves in the Bohai and Yellow Seas in the north, and the South China Sea in the south. Although a number of small to medium-sized fields have been discovered

TABLE 5.12. *Stages in Sino-foreign co-operation in offshore oil*[a,b]

Stage	Years	Principal characteristics
1	1979–80	Geophysical reconnaissance surveys carried out in a 420,000 sq. km area in South China Sea and southern part of Yellow Sea.
2	1980–2	Five petroleum contracts signed through bilateral negotiations with foreign oil companies. Contracts covered selected areas in Bohai Sea, the Beibu Gulf, and the Yinggehai Basin.
3	1982–4	Promulgation in 1982 of the Regulations on the Exploitation of Offshore Petroleum Resources in Cooperation with Foreign Enterprises and other decrees. Nineteen petroleum exploration contracts awarded to foreign oil companies during first round of bidding.
4	1984–9	Second round of bidding initiated in November 1984. Nine contract areas were taken up during the second round. In 1986, despite slumping oil prices, five oil contracts were signed with foreign companies plus a geophysical prospecting agreement.
5	1989	The third round of bidding lasted just four months, January–April, though bilateral negotiations continued past the closing of the round. To attract greater foreign interest, inducements (flexible negotiations) and incentives (new royalty rates) were promoted by CNOOC and the Ministry of Finance. Seven bidding blocks were offered in the round in the Pearl River Mouth Basin.
6	1992–3	The fourth round of bidding was initiated 30 June 1992 by CNOOC. It was scheduled for completion by July 1993, with exploration due to start in 1994. The offshore area for bidding lies in the East China Sea and covers a total acreage of 72,800 sq. km. Foreign firms' bids were expected to focus on two large oil-bearing areas.

[a] By June 1992 CNOOC had signed 70 contracts and agreements with 50 oil firms from 13 countries on petroleum exploration and development. More than half a million sq. km of offshore areas in the South China Sea, Bohai Sea, and East China Sea have been opened to foreign enterprises.

[b] Between 1979 and 1992, more than $4 bn. has been invested by foreign companies in the search for oil off the coast of China.

Sources: *China Daily* (1 July 1992); Wu Kang (1989); Zhu Ang (1988); Gorst (1988).

in various geologic depressions, most of the finds have been disappointingly small. To date, more than 500,000 square kilometres of offshore areas in the South China Sea and the Bohai Sea have been opened to foreign firms (*China Daily*, 1 July 1992). Six offshore fields are producing currently at a rate of 6,600 tonnes/day. They are Chengbei oil field, BZ 28-1 oil field, BZ 34-2/4 oil field, and 426W oil field in the Bohai Gulf under the supervision of the Bohai Oil Corporation, Wei 10-3 oil field in the Beibu Gulf under the supervision of the Nanhai West Oil Corp., and Huizhou 21-1 oil field under the supervision of the Nanhai East Oil Corporation (China National Offshore Oil Corporation 1992).

Sino-foreign co-operation in Chinese offshore oil exploration and development has passed through six major stages since its initiation in 1979 (Table 5.12). These have included the initial reconnaissance surveys carried out over the course of about one year, three completed rounds of bidding, and a fourth round scheduled for completion in July 1993. In addition to bidding in official rounds, foreign countries can also negotiate individual exploration contracts with Chinese officials. Such bilateral negotiations are continuing today. Only nine contract areas were taken up in the second round, compared with 19 in the first round. Third-round bidding closed in April 1989 and was generally lacklustre. Seven contract areas were offered covering more than 32,000 square kilometres in the Dongsha-Shenhu region to the south of the Pearl River Mouth discoveries (Gorst 1989). By the end of 1989 one contract was formally signed between the China National Offshore Oil Company (CNOOC) and Occidental Petroleum for exploration rights.

To attract greater interest in the much delayed third round of bidding, CNOOC adopted a more flexible approach in the negotiations for exploration contracts, and fiscal terms were relaxed. Government officials also reduced state fees on smaller oil (and gas) fields to encourage foreign investors. After the third round of bidding closed, CNOOC announced that 14 foreign companies requested bidding documents, with 12 of them having participated in earlier rounds. There were two new companies participating in the round—Mobil and Murphy Oil.

The fourth round of bidding was announced in Beijing on 30 June 1992 by the Executive Vice-President of CNOOC. Twenty-four international oil companies were on hand during the announcement to be briefed on oil prospects in the previously closed East China Sea. The two areas for bidding were 18,800 square kilometres and 54,000 square kilometres respectively, consisting of 20 blocks. Waters are 50 to 100 metres deep in the areas. The opening of the East China Sea in the fourth round means Sino-foreign co-operation now includes all of the seas surrounding China's mainland.

Since much of China's offshore oil discoveries are marginal at \$15/barrel–\$20/barrel and subcommercial below \$15/barrel, future levels of foreign investment will be largely influenced by world oil prices (Woodard 1987). Another factor to be considered is the possible discovery of any major field with commercially viable oil and/or gas. One of the most important offshore discoveries to date has been Bohai Oil Corporation's find in Liaodong Bay, which exhibited China's ability to discover offshore oil without foreign assistance. Chinese government officials do openly admit, however, that the country requires foreign assistance to promote offshore oil development and share risks for petroleum exploration.

In terms of foreign investor interest in China's onshore exploration and development, government officials are now beginning to ease restrictions on investments into the nation's interior provinces. In 1993, for example, China began or expanded efforts to attract foreign capital into oil development projects within Hebei, Hubei, Liaoning, and Shandong provinces, Tianjin Municipality, and the Xinjiang Uygur Autonomous Region. China also opened three areas of the oil-rich, but largely unexplored, Tarim Basin of Xinjiang to foreign firms able to provide advanced technology and investment. The Tarim Basin is believed to hold approximately one-seventh of China's oil reserves.

If world oil prices rise towards the mid-1990s as expected by many analysts, investments into China's onshore oil industry will be likely to pick up, with significant capital outlays possible in the large and highly prospective basins of the remote western provinces where interest is greatest. SINOPEC recently established a joint venture in Tianjin with US-based Honeywell, Inc. to develop technology systems, including both software and hardware, for the hydrocarbon processing industry and market (*China Daily*, 17 Jan. 1993). SINOPEC–Honeywell (Tianjin) Ltd. is Honeywell's first joint venture in China; it has a total investment of more than \$4m. The majority of the assembled systems will be marketed in China to meet the country's rising demand for advanced control and automation systems.

An important development of the Seventh Five-Year Plan period was the gradual opening of China's refining sector to foreign involvement. Foreign-funded projects are considered a key to the continued expansion of the Chinese refinery industry in the 1990s. In July 1992 Royal Dutch-Shell initiated foreign interest in refining by proposing a \$2.5bn., 100,000 b/d refinery/petrochemical plant to be built in Guangdong province (Fridley 1991*a*). The agreement to undertake a feasibility study for this project was signed in Beijing on 29 July by Shell and five Chinese partners (who will provide 50 per cent of the investment), including China National Offshore Oil Corporation (CNOOC), CNPC, SINOPEC, the Guangdong provincial government, and China Merchant Holding Company. If the project ultimately moves forward, the parties will sign a contract to form one of the largest joint ventures in China. Other foreign ventures in refining have been proposed at both Dalian and Shanghai. A new 60,000 b/d refinery serving Shenzhen and Hong Kong is planned for Dapeng Bay in Shenzhen, financed by the Shenzhen Municipal Government, SINOPEC, and China Resources of Hong Kong.

Another energy resource, namely coal, was the focal commodity of one of the largest Sino-foreign joint ventures in China during the 1980s—the Occidental Antaibao coal project. Located in Shanxi province, this project represented a \$700m. investment, with Occidental contributing an estimated \$345m. (Zhu Ang 1988: 136). With the death of Armand Hammer, the long-time Occidental Petroleum Corporation Chairman, in December 1990, Occidental announced its pull-out on 14 January 1991 as part of worldwide asset sales aimed at slashing the company's debt. Occidental's departure came at a time when the mine was beginning to operate according to plan, although in 1990, its best year in terms of output, Antaibao produced at 75 per cent of its 12m. tonne capacity, with exports at less than half of the 8–9m. tonnes originally forecast by Occidental.

The Antaibao mine, one of the largest surface mines in the world, opened in September 1987 after years of frustrating planning and negotiations. The project was pioneering in its development, financing, and size, owing largely to the personal commitment and perseverance of Occidental's Armand Hammer and Chinese Supreme Leader Deng Xiaoping, who elevated the project to a symbol of China's newly implemented open-door policy. Many of Antaibao's problems surprisingly resulted from expanding economic reform in the country, as Beijing's weakening centralized control on the economy has led to complications in the provision of infrastructure and power to the project once done automatically.

Ties between Occidental Petroleum Corporation and China will continue at some level despite the company's departure from Antaibao. Improved cash flows resulting from the groups' asset sales may be used to support Oxychem, the company's chemicals division operating in China. In 1991 Oxychem was considering its first equity joint venture involving a modest $10m. investment in an existing raw materials factory in Shandong province (*Far Eastern Economic Review*, 7 Feb. 1991).

Aside from Occidental's involvement at the Antaibao mine, the only other major foreign activity in China's coal industry was a Japanese concession loan in 1989 of $800m. to facilitate development of the Jungar project in Inner Mongolia. Direct foreign investment in Chinese coal operations is relatively rare as the government remains cautious in relinquishing control over the sector and profit margins are generally small in the industry.

As with energy, China's minerals industry has also expanded ties with foreign countries and companies in an attempt to promote development. Foreign investment is one of the major financial resources available to the industry—which is in desperate need of investment owing to limited domestic financing, an increasing reliance on imports, and a growing demand for metals. China's Eighth Five-Year Plan (1991–5) calls for an average annual growth rate in GNP of 6 per cent, with industrial output value in 1995 rising to more than 37 per cent the 1990 level (*Beijing Review*, 13–19 May 1991). China's ambitious economic and industrial targets imply continued reliance on the expansion of the minerals sector. This, in turn, suggests greater need for foreign financial and technical assistance in the years ahead.

While China's mining industry is today one of the world's largest, it is plagued by numerous problems which may threaten future expansion, including low efficiencies in recovery, production, and utilization, outdated mining equipment and technology, and severe transportation bottlenecks. Billions of dollars of investment will be required to facilitate the industry during the 1990s. The availability to capital into priority investment areas will be critical in order to ensure expansion of the industry into the twenty-first century. Priority will be given to the development of mines with large volumes and high-grade deposits, low production costs and economically viable operations. Stress will be placed on comprehensive prospecting, mining, recovery, and utilization, and the intensive processing of manufactures for the purpose of absorbing foreign investment. The primary investment areas in China's mining industry include:

- metal smelters and refineries (principally for copper, tin, tungsten, aluminium, lead/zinc, rare earths, and nickel);
- specialized mining and processing software and excavation techniques;
- geological exploration technology;
- modernizing iron and steel plants;
- new and used mining and drilling machinery;
- pollution monitoring and controlling devices;
- development of mineral sands for export;
- engineering, computer, and consultancy services; and
- physical infrastructure, transportation, and telecommunications.

Utilizing imported technology and equipment to renovate existing enterprises is a Chinese strategy to tap enterprise potential and promote production, economic growth, and technical development. During the 1990s and early part of next century technology transfers could play a pivotal role in promoting domestic economic growth as well as co-operation with foreign nations. Though economic restructuring is under way in China, technological advancement of industry is needed to facilitate economic growth in the long term. It has been estimated that as much as 75 per cent of all equipment used in Chinese industry must be renovated or replaced to improve efficiencies in operations (Dorian 1987). Such an accomplishment would require massive capital investments as well as technology transfer.

Potential growth areas for imports of technology and equipment are in priority development projects, which include minerals and energy extraction, processing, mining and safety, and pollution abatement equipment. China is intent on expanding its minerals and metals production in order to manufacture value-added goods to meet consumer needs. Production of minerals, metals, and their products will also depend on solving ecological problems.

In China, production of mineral commodities is regarded as a means of facilitating industrial expansion and an avenue for promoting economic development in sparsely populated areas, such as the Xinjiang autonomous region. Selected minerals are used for export to earn hard currency and to fulfil barter trade arrangements. Historically, Chinese policies concerning minerals were based largely on self-sufficiency motives. With the advent of the open-door policies, however, the Chinese government recast its national mineral policy to coincide with economic and political restructuring. China has fully realized that expansion of its minerals industry cannot take place without a substantial influx of technology, science, and capital from abroad, which is likely to occur through the fostering of joint ventures.

Though the Chinese government is actively seeking overseas partners in mining-related joint ventures, foreign mining companies have adopted a cautious approach in entering the Chinese market, as in other fields. Major mining companies that had sought investments in the 1980s in the minerals field were generally disappointed with the deposits offered to foreign investors, a lack of reliable geologic and economic data, and Chinese bureaucratic complications. Moreover, investments in the mineral sectors suffer from the same general problems as other foreign investment in China, notably difficulties in obtaining foreign exchange to repatriate profits, the rising cost of materials and wages, shortages and limited access to needed materials, inadequate infrastructure, a largely untested and fragile legal system, and uncertainty about consistency in government policies and procedures. Up to 1992, only a handful of joint ventures in the minerals field were registered with the Chinese government, though additional projects were under negotiation.

During the 1980s, most foreign involvement—whether in the form of investment, joint-venture activities, or technology transfers—was used to boost production of aluminium, copper, and lead and zinc. Foreign companies will continue being encouraged to participate in the development of China's domestic metals industry through sales of mining and plant equipment, engineering services, computers and controls, and technology transfers.

The principal geographic area for attracting foreign capital in China's non-ferrous metals industry is south-eastern China and, to a lesser extent, the north-west. Because of geologic conditions, these regions have abundant metal resources and large concentrations of high-grade lead, zinc, and bauxite, as well as promising hydroelectric power potential. They offer relatively attractive environments for joint-venture activities, with prospects of low investment and relatively rapid financial returns. Co-operation with foreign enterprises on non-ferrous metals production in China's many coastal cities is also encouraged.

Table 5.13 lists the major projects involving foreign companies in China's metals industry between 1989 and 1992. The profiles are based on media reports of business contracts and negotiations. The table is not necessarily complete, though it does provide an accurate representation of the types of projects involving foreign

TABLE 5.13. *A profile of major investment projects in China's metals and minerals industry, 1989–1992*[a]

ALUMINIUM

Link Hing Metal Co. (Hong Kong)/Nanhai Dali Urban Economic Development Co.

Asian Aluminum Factory joint venture in Guangdong Province began operations. Est. annual production capacity is 12,000 tonnes. Jan. 1992

Alcan Aluminum Ltd. (Canada) and Japanese Light Metal Corp. (Japan)/CNNC, Southwest Aluminum Processing Factory, and Guizhou Aluminum Factory

Huajia Aluminum Materials Co. Ltd. joint venture in Shenzhen began production of aluminium products. $23.2m. (Canada and Japan—45%, China—55%). Aug. 1990.

TABLE 5.13. (*cont.*)

ABC Co. (USA)/Nanjing Aluminum Products Factory
Established Nanjing–ABC Aluminum Products Co. Ltd. to produce aluminium and stainless products. $200,000. (USA—25%, China—75%). Oct. 1989.

NA (UK)/Fenghua Machinery Co. and Chemical Reagent Plant, Heilongjiang
Established Heilongjiang Licheng Aluminium Foil Paper Products Co. Ltd. to produce 120m. sq. m aluminium foil p.a. $9.2m. Feb. 1989.

Mino (Italy)
Negotiating for construction of aluminium plant. $115m. (IL 150bn.). Jan. 1989.

COAL

Corub Kopes Co. (FRG)/Anshan Coking and Refractory Material Engineering Co.
Will establish Sino-German Engineering Co. for coking. Mar. 1989.

Nissho Iwai Corp. (Japan) and JGC Corp. (Japan)/Yanzhou Coal Mine Administration, Shandong
Established joint venture to produce coal and water mixture. $24.9m. (yen 3.8bn.). (Japan—49%, China—51%). Registered capital: $8.3m. (yen 1.28bn.). Jan. 1990.

CONSTRUCTION STONE

Top Winner Co. Ltd. (Hong Kong)/Four Chinese companies
Opened the Jinxin Granite and Marble Co. Ltd. in Tianjin. $3.68m. (yen 13.7m.). May 1989.

COPPER

Philipp and Lion Ltd. (UK)/CNNC Shanghai Smelter and Cable Wire Group of Shanghai (CGS)
Signed letter of intent for a joint venture to construct a modern, economically efficient pollution-free copper smelter in Wusong Industrial Area, Shanghai. (UK—50%, China—50%). Mar. 1991.

United Trading Co. Ltd. (Canada)/Cangzhou, Hebei
Established Cangyi Welding Wire Co. 10-year joint venture to produce copper-plated welding wire. $1.8m. (Canada—60%, China—40%). Oct. 1989.

Wrought Copper Ltd. (Chile)/No. 1 Copper Tube Plant, Beijing
Started operation of the Jingsheng Copper Tube Co. Ltd. joint venture to produce red copper tube. $10.64m. (50%–50%). May 1989.

NA (USA)
Established Jinhui Reprocessed Metals Co. Ltd. to process copper and aluminium ingots. $580,000. Apr. 1989.

GEMSTONES

Argyle Diamonds Corp. (Australia)/Government of China, Beijing[b]
Argyle has opened a diamond-cutting and polishing factory about 30 km from Beijing. The factory will provide training for Chinese in the crafts of cutting and polishing diamonds. Sept. 1991.

Pacific Gem Co. Ltd. (Australia)/Changle County, Shandong
Signed agreement to jointly process sapphires. $1m. June 1989.

TABLE 5.13. (*cont.*)

GOLD

Galactic Resources Ltd. (Canada)
Assisting in gold-mining project. Dec. 1988.

IRON AND STEEL

Asian Development Bank (ADB)
Will provide loan to modernize and expand the Laiwu Iron and Steel mill in Shandong province. $133m. Apr. 1992.

Fried. Krupp GmbH (FRG)/Talyuan Iron and Steel Co. and CITIC
Established joint venture in Taiyuan, Shanxi province, to produce steel. $400m. (FRG—30%, China—70%). Sept. 1990.

Australia
Established the China–Australia Iron and Steel Industry Training Center at the Wuhan Institute of Iron and Steel Engineering to train management personnel and engineers over a five-year period. $10.26m. (Australia—67%, China—33%). Sept. 1990.

Nittetsu Shoji Co., subsidiary of Nippon Steel Corp. (Japan) and Mitsui and Co. (Japan)/Shum Yip Trading Co. and Nanhai Oil Shenzhen Development and Service Corp.
Begin operation of joint-venture steel-cutting plant in Shenzhen special economic zone. $8m. (Japan—65%, China—35%). Mar. 1990.

Mitsubishi Heavy Industries Ltd. (Japan) and Sumitomo Metal Industries Ltd. (Japan)/Ministry of Metallurgical Industry
Will establish joint venture to design and engineer continuous steel-casting machinery. Jan. 1990.

Mesta Engineering Co. (USA—70% owned by Capital Iron and Steel Corp.)/Capital Iron and Steel Corp., Beijing
Opened Beijing Mesta Engineering Co. Ltd. joint venture to upgrade steel production process and explore foreign markets. Sept. 1989.

NA (USA)
Established Yichang Steel Strip Co. Ltd. 15-year joint venture to produce thin steel and tin-plated strips. $30m. Sept. 1989.

International Iron and Steel Far East Service Co. (USA)
Will set up plant at Wuhan Iron and Steel Works to process steel slag and recover scrap steel. $7.5m. July 1989.

Always King Ltd. (Hong Kong)/Shanxi Energy Corp. and Yuanping Iron and Steel Works
Signed agreement to build 150 cu. m puddling furnace and 24 cu. m sintering machine. May 1989.

NA (Japan)/Baoshan Steel Works, Shanghai
Established Baohua Trading Corp. to deal in steel, non-ferrous metals, mechanical equipment, etc. Registered capital: $340,000 (50%–50%). Jan. 1989.

Scandiaconsult (Sweden)
Preparing feasibility study for iron mine and iron ore dressing plant. Jan. 1989.

Mestek Inc. (USA)/Shoudu Iron and Steel
Sold 19% of Mesta Engineering Co. to Shoudu, the majority partner in Mesta. (USA—30%, China—70%). Jan. 1989.

TABLE 5.13. (*cont.*)

METALLURGY

NA (Hong Kong)/Tianjin
Established Huasheng Co. to process metal waste. Registered capital: $400,000. Nov. 1988.

RARE EARTHS

NA (Canada)/Jiangxi
Established Gongjia Rare Earth Co. Ltd. joint venture to process 200 t/y rare earths. Sept. 1989.

NA (Hong Kong)/Yongfu County, Guangxi
Established Guiyong Magnetic Material Plant to mine and process nafeb alloy, a rare earth magnetic material. Feb. 1989.

NA (Hong Kong)/Shanghai Rare Metals Refinery and Machinery Installation Co., affiliate of China No. 1 Metallurgical Construction Corp.
Established Shenli Rare Metals Co. Ltd. joint venture to recycle and refine rare metals. July 1989.

TALC

Ilshin Industrial (Rep. of Korea) and Cyprus (USA)
Will develop talc mines. Mar. 1989.

TIN

Malaysia Mining Corp./CNIEC
Will form joint venture to mine tin. Jan. 1989.

TITANIUM

Marubeni Corp. (Japan) and Malaysia Mining Corp. (Malaysia)/Hainan Provincial Metallurgical Nonferrous Metal Industry Corp.
Established joint venture in Hainan province to mine titanium oxide-bearing mineral ilmenite for export to Japan. Funds involved est. 100m. yen. July 1990.

General Minerals Co. (Malaysia)/Yunnan Geological and Mineral Resources Bureau
Signed letter of intent to exploit titanium-containing ores. Jan. 1989.

Notes: NA not available. CITIC China International Trust and Investment Corp. CNIEC China National Nonferrous Import and Export Corp. CNNC China National Nonferrous Metals Industry Corp. CGS Cable and Wire Group of Shanghai.

[a] Based on press reports of business contracts and negotiations. Not necessarily a complete list or independently verified. Contracts denominated in foreign currencies were converted into $US at the most recent monthly average rate quoted in International Financial Statistics (IMF).

[b] Set up as a joint project between the governments of Australia and China under the Technical Cooperation Agreement established in 1991 for bilateral aid.

Sources: US–China Business Council (1989–92: various issues); *Mining Journal* (1989–92: various issues); Dorian (1992); author's files.

firms, the range of commodities attracting interest, and the countries represented in the deals. Where available, investment figures are included in the profiles.

As shown, a majority of Sino-foreign projects are in aluminium, copper, and iron and steel. Other commodities attracting overseas investment include coal (Japan, Germany), construction stone (Hong Kong), gemstones (Australia), gold (Canada), rare earths (Canada, Hong Kong), talc (South Korea, United States), tin (Malaysia), and titanium (Japan, Malaysia). Twelve projects involving iron and steel production and processing were under negotiation or finalized during the 1989–92 period, involving Germany, Australia, Japan, the United States, Hong Kong, and Sweden. Altogether 13 countries are represented in the investment projects listed in Table 5.13.

Although only one project involving joint gold development is documented in Table 5.13, China's gold industry is determined to strengthen co-operation and technical exchanges with foreign investors in a bid to accelerate its development. At a June 1993 symposium, the Chinese Vice-Minister of the Ministry of Metallurgical Industry and Director of the State Gold Administration announced that the government has approved foreign investment in the nation's gold industry, which has been severely off limits to overseas investment for more than four decades (*China Daily*, 16 June 1993). However, foreign investment will be limited to those gold deposits which are too technically difficult for China to mine on its own. Such mines are located in Shandong, Hebei, and Shaanxi provinces.

In what was dubbed the first base metals exploration joint venture between a foreign company and a Chinese partner, Australian-based Broken Hill Proprietary Company (BHP) signed an agreement in January 1993 to search for zinc in China's Sichuan province (*Mining Journal*, 22 Jan. 1993). The agreement is between BHP's minerals division, the provincial-level enterprise Sichuan Non-ferrous Metals Ore Resources Development Corporation (SREDCO), and China's Southwest Metallurgical and Geological Exploration Company, part of the Ministry of Metallurgical Industry. BHP Minerals had been pursuing joint-venture opportunities in China since 1986, and the January accord fulfilled a letter of intent signed by the parties in October 1991.

3. *China's Asia–Pacific Role in Minerals and Energy*

Owing to China's open-door policies and growth in its mineral and energy production and trade, the nation is playing an increasingly prominent regional role in these sectors. This role is one of both a major exporter (tungsten, tin, rare-earth oxides, oil, and coal) and importer of raw materials (iron ore, aluminium, copper, and several non-metal commodities). China is also competing for international capital through borrowing or foreign investment to develop its indigenous resources. As China's participation in the mineral and energy economy of the Asia–Pacific region grows and changes in character, new issues arise that affect China's relations with its neighbours. Three important issues that are likely to grow in prominence in the 1990s are China's role as a competitor, China's potential role as an investor in regional resource development, and the political implications of mineral and energy trade for China's relations with its neighbours (Clark, Dorian, and Fridley 1989).

As China initiated policies after 1978 to sell mineral and energy products more aggressively and attract foreign investment, it was immediately regarded by other resource-exporting Asia–Pacific nations as a formidable competitor. The Indonesian government, for example, was deeply concerned and made strong presentations to the Japanese that more competitively priced Chinese crude petroleum was displacing purchases of Indonesian crude in the Japanese market. The Association of Southeast Asian Nations (ASEAN) group as a whole expressed its concern to the Japanese government that its foreign assistance and loans to China should not come at the expense of the ASEAN countries.

China will continue to play a role as a competitor in important resource and investment areas. For example, as it seeks to replace declining foreign exchange revenues from petroleum with increased revenues from coal exports, its coal is competing directly with Australian coal. However, China's role as a competitor is changing in two important ways. First, with a growing demand that cannot be met by increased domestic supplies, it will be competing increasingly for access to foreign mineral and energy supplies. Secondly, in aspiring to capture the value added to resources by processing, it will be competing with both resource-rich countries and more advanced economies that have established processing industries.

The shift in China's position from an export competitor to a competitor for supplies is likely to be felt first in the petroleum industry. Within a few years, China's exports of crude petroleum and petroleum products will decline to only a quarter to a third of its present level while its imports of petroleum products could increase dramatically. In particular, China is likely to become a major importer of diesel fuel in a regional market that is already very tight.

China's efforts to establish itself as a regional processing centre may have an even greater impact. Chinese policymakers are aware that their country loses potential revenue and jobs by exporting raw materials rather than processing or refining products. For imported products, China can reduce its costs by importing raw materials and processing them at home. In some cases, China already has surplus processing capacity, in part because the processing facilities are located at uneconomical distances from supply sources in China. Moreover, with relatively low labour and energy costs, China may have comparative advantage in resource processing.

As China's resource-consuming industries develop, the nation will seek to ensure security of raw material supplies abroad by investing in mining or processing industries. In a major step of this kind, the China International Trust and Investment Corporation (CITIC) invested $120m. in the Portland Aluminium Smelter in Victoria, Australia, in 1986. This investment included a 10 per cent equity participation and entitled China to 10 per cent of the output. The smelter is a joint venture 45 per cent owned by Alcoa, 35 per cent by the State of Victoria, 10 per cent by the First National Trust, and 10 per cent by CITIC. In 1993 CITIC announced its intention to increase its 10 per cent stake in the Portland Aluminium Smelter, which presently has an electrolytic aluminium capacity of 300,000 tonnes per year (*Mining Journal*, 15 Jan. 1993).

In addition to aluminium, China's CITIC group is today also considering investments in Australian coal, copper, and gold, and is reportedly holding discussions with relevant company and government officials. China has also undertaken copper investments in Papua New Guinea, similarly allowing the nation to acquire a claim over foreign resources for products in short supply in China.

Overseas investments may occur in the petroleum industry for the same reason or to protect China's market share. For example, China responded to a tightening of US lead standards that resulted in a loss of exports to the United States from China by investing directly in a refining and marketing company in California. SINOPEC, described in Chapter 4, has bid on petrochemical projects in Malaysia to secure a claim on the products. If China becomes a net importer of crude petroleum because of increased demand at home, it may, like Japan, the Republic of Korea, or Taiwan, adopt a strategy of investing in overseas exploration and development activities.

In terms of relations with complementary economies, China's mineral and energy trade may actually play a political role in helping improve them. Sino-Japanese political relations in the 1970s and early 1980s were, for example, reinforced by the strong complementary needs of the two economies; Japan was interested in Chinese petroleum and coal and China needed Japanese capital goods, investment, and technology. In the 1990s, such economic ties could be a further stimulus for improvement in mainland China's evolving relations with the former Soviet republics, the Republic of Korea, and Taiwan. China is a major producer of metals required by these neighbouring economies.

For example, the Xinjiang Uygur Autonomous Region in north-west China and Heilongjiang province in the north-east are believed to be rich in minerals but have been poorly explored and developed. Both areas border the former Soviet Union and have recently improved transportation corridors to both Russia and Kazakhstan to facilitate cross-border trade. Mineral raw materials in these regions could be sold or bartered to the former Soviet republics and perhaps processed more efficiently there than in China. The prospects for Sino-Russian trade in energy resources are less favourable. Even if China becomes more interested in purchasing crude petroleum, Russian crude generally has a high sulphur content. China would have to invest in expensive desulphurization equipment to use it. However, there are good prospects for border trade in petroleum products in short supply locally.

The Republic of Korea and Taiwan, like Japan, are large markets for low-sulphur crudes, fuel oils, and coal. The Republic of Korea has already begun to import Chinese oil and coal, capturing the economies of the short delivery distance between the two countries. Korean imports of coal increased rapidly over the past decade from 7m. tonnes in 1980 to 22m. tonnes in 1990, and are expected to rise to 48m. and 65m. tonnes in 2000 and 2010 respectively (Lee 1991).

With renewed emphasis in Taiwan on coal-fired power plants, especially with the shelving of its proposed fourth nuclear plant, coal imports by Taiwan have doubled in the past few years. Taiwan has very limited energy resources, and is heavily dependent on imported energy. Coal's share of total energy supplies is projected to increase from 24 per cent in 1990 to 31 per cent in 2000 (Johnson 1992). In 1988, Occidental Petroleum offered to sell coal from its Antaibao mine to Taiwan, but Taiwan's leadership shunned the proposal for the time being. Nevertheless, Taiwan recently permitted the indirect sale of raw materials from the mainland, and the first shipment of mainland coal arrived in October 1988. Taiwan and mainland Chinese officials are reported to have started discussions in Hong Kong over possible mainland oil sales to Taiwan.

As the Korean and Taiwanese currencies appreciate, these two economies will be seeking the cheapest possible sources of raw materials and processed inputs to maintain their competitiveness. Since China is a low-cost producer of such goods, neither the Republic of Korea nor Taiwan will be able to ignore the opportunities for increased trade with China. In addition, China does offer promising investment opportunities to the Republic of Korea and Taiwan—two countries which are already major investors in the People's Republic.

CONCLUSIONS: FOREIGN ECONOMIC RELATIONS

Until the 1980s, China's foreign trade and economic relations with other countries played only a minor role in the nation's development. The relatively small level of trade that occurred was focused on the import of domestic deficiencies and the export of surpluses. Following the Third Plenary Session of the 11th Party Central Committee in 1978, the opening of China to international exchanges became an important feature of the country's economic strategy. Foreign trade, investment, technology assistance, and joint-venture participation were to help China achieve its ambitious economic and industrial goals.

China's movement into the international economy can be measured by the expansion of total trade. The total volume of China's foreign trade soared from \$29.3bn. in 1979 to \$135.7bn. in 1991, or 32.7 per cent of the gross national product. By 1992, China had utilized foreign loans valued at more than \$50bn. Since 1979, more than 40,000 foreign-funded business enterprises have been established in China, with more than half of them involving joint-venture participation. Statistics indicate that more than 25 per cent of China's total foreign trade in 1992 was generated by enterprises with foreign investment (*China Daily*, 9 Jan. 1993).

Minerals and energy trade accounts for about 14 per cent of China's total trade turnover, with exports estimated at $11.08bn. in 1991, imports at $7.71bn. China's minerals and energy trade balance was positive in 1991, showing a surplus of $3.37bn. Notably, however, the export value of petroleum and petroleum products has declined since 1985, and imports of the goods are rising. By the mid-1990s China may become a net importer of petroleum.

To further develop China's mining industry, some level of foreign assistance in the form of trade, investment, technology transfer, equipment sales, and engineering services will be required. Except for offshore oil exploration and development, foreign commitments to participate in China's mining development have been rather small in number. For this reason government officials continue to stress the opening of the economy to international exchanges, and are gradually facilitating increased foreign investments into such areas as petroleum downstream refining.

China's enormous geological potential may eventually offer substantial opportunities for joint-venture investment, particularly in non-ferrous metals extraction and processing, recovery and utilization of base metals, conservation of energy, and oil exploration, drilling, and enhanced recovery. As China strives to increase its foreign investment base in the 1990s, the principal investment partners will continue to be Hong Kong, Taiwan, Singapore, the Republic of Korea, and Japan, as well as Australia, the United States, and selected European countries. Companies interested in pursuing business opportunities in China's mining industry will need to remain flexible and patient, and attempt to understand not only the industry, but the often confusing bureaucratic systems. The planned expansion and modernization of China's minerals sector will require a tremendous influx of foreign technology and financing. Firms that are able to establish a successful track record today in China may be able to reap more substantial rewards in the long term.

NOTES

1. According to Fridley (1990), China will probably import more than 200,000 b/d of naphtha and diesel by the mid-1990s. If the goal of 165m. tonnes (3.3m. b/d) of refining capacity by 1995 is reached, then crude processing levels may be high enough to keep product imports relatively low, but at any lower levels of throughput imports will constitute a significant fraction of domestic consumption.
2. Based on a China coal trade model developed in 1988 at the East–West Center, Honolulu, Hawaii. See Johnson (1992) for discussion of China's coal future to 2010.
3. Direct foreign investment (DFI) is more broadly defined in Chinese than in Western usage. Under Chinese practice, DFI includes investment in equity, joint ventures and wholly owned ventures (as the term is generally understood outside of China), as well as

co-operative arrangements with the participation of foreign capital, counter trade, and the like. See US Department of Commerce (1988: 64).

4. In 1991 China granted Papua New Guinea a $10m. interest-free loan for development projects. The agreement provides for both countries to explore opportunities to enhance trade and investment in primary and secondary produce including copper concentrates.

6
Mining and the Chinese Economy

MODELLING MINERAL DEVELOPMENT IN THE CHINESE ECONOMY

A number of studies (for example, Torries, Rose, and Chen 1988; Pintz and Rizer 1985; Johnson and Pintz 1985; Radetzki 1982; and Nankani 1979) have discussed the development impacts of mining and mineral processing on the regions in which they are located. The impact of mining in China is felt throughout the nation's economy in terms of employment and income generation, and the production of raw materials for use in the manufacture of semi-finished and finished products (see Table 6.1). Mining projects in the country create large but short-term demands for employment, capital, and raw-material inputs, and the effects of such demands are transmitted by the minerals sector to the rest of the economy.

An important question about the mining sector of China concerns the magnitude of the sector's economic impact on national output, income, and employment. The mining sector purchases products from other sectors for use in its production and investment activities. Therefore, by way of such backward linkages, China's mining sector directly influences output, income, and employment in those sectors which supply its inputs. In the same line, the mining sector also indirectly influences the sectors that supply its suppliers, and so on. At the same time, if an increase in domestic sectoral output occurs, this additional output may serve as inputs to other industries and undergo further processing and refining (forward linkage effect). Hence, through such intersectoral sales and purchases in the Chinese economy, an initial change in the mining sector has a widespread secondary influence.

One way to systematically analyse the mutual interdependencies between the different parts of an economy is to use the input–output technique. This technique of economic analysis describes an economy as a system of interdependent activities—both in a direct and in an indirect manner. Hence it is particularly suited to a detailed quantitative assessment of both the direct and indirect secondary effects of any single policy or combination of policies, although the extent of assessment by input–output analysis is largely dependent on the data available. Input–output analysis is essentially a linear model of purchases and sales (transactions) of all of the sectors of an economy based upon the technological relationships of production. Because of its general equilibrium character, the input–output technique has been recognized as a useful method for analysing overall economic impacts.

TABLE 6.1. *Possible economic impacts[a] of a mining venture in China*

Economic Impacts	National (macro-assessment)		Sub-national (regional/sub-regional)		Project site (local community)	
	Direct	Indirect	Direct	Indirect	Direct	Indirect
Employment	×	×	×	×	×	×
Revenue	×	×			×	×
Foreign exchange earnings	×	×				
Export diversification	×					
Population movements		×	×	×	×	×
Environmental	×		×	×	×	×
Training	×	×	×	×	×	×
Local business development		×		×	×	×
New industries		×		×	×	×
Urban planning/housing			×	×	×	×
Infrastructure			×	×	×	
Education				×	×	×
Traditional agriculture					×	×

[a] Impacts can be categorized as either direct or indirect. Direct impacts are related directly to the mining venture. Indirect impacts include (1) backward linkages (through requirement of inputs), (2) forward linkages (much more extended than backward chain in mining), and (3) other linkages (related to infrastructural facilities which form part of the minerals venture).

Sources: Pintz and Rizer (1985); Radetzki (1982); Jackson *et al.* (1983).

The usefulness of the input–output technique extends to centrally planned economies. Input–output modelling offers central planners the possibility of assuring the internal consistency of an economic plan, that is, seeing that the pre-assigned output targets are consistent with the input requirements. When studying Soviet economic planning in the 1960s, Levine (1962) concluded that input–output analysis of a planned economy offers numerous advantages, but also some disadvantages as well.[1] An important difference between the application of input–output modelling in a free-market economy and in a centrally planned economy is that with a free-market economy the analysis will begin with a set of final demands, whereas in a centrally planned economy the targets established are not final demands but total outputs (Miernyk 1965: 87). Total ouputs include not only the final demands but also the inter-industry transactions needed to achieve these final uses.

With respect to China, Western understanding of intersectoral transactions has long been impeded by the unavailability of input–output tables for that country. Polenske (1986) has documented the development of input–output research in China, in terms of both national and regional input–output analysis. Until 1987 acquiring input–output tables produced in China was difficult owing to the sensitivity of the economic data contained within them. However, in 1987, 1981 input–output tables of China (in producers' value[2]) were published by the China Statistical Information and Consultancy Service Centre of Beijing, in conjunction with the East–West Center of Honolulu, Hawaii. More recently, detailed regional and provincial input–output tables have been released. Thus, for the first time, Western researchers have access to published input–output tables of China permitting analyses of the nation's economic and industrial structure. The 1981 tables were originally compiled by China's Centre of Economic Forecasting in collaboration with other concerned departments of the State Council.[3]

The newly released tables were compiled in two manners: by the industry method and the commodity method. The industry-by-industry input–output table is useful in analysing the efficiencies in industrial output changes over time. The commodity method of analysis is based on the principle of homogeneity, where products are homogeneous in respect to economic usage, technology, or structure of intermediate consumption. Although a commodity-by-commodity table is generally considered more useful for analysis by economists, the commodity approach is inconsistent with the method statistics for commodity production that are collected in China (China Statistical Information and Consultancy Service Centre and the East–West Center 1987). For the purposes of this study, the industry-by-industry input–output table was employed.

The table disaggregates the Chinese economy into 24 material product sectors. Input–output data were compiled by industry sectors where the gross output value of an enterprise is classified into the industry to which the predominant products of the enterprise belong. For example, the value of machinery produced by a metallurgical enterprise is included in the industry of metallurgy. Such treatment is consistent with the prevailing economic accounting system in China and, in general, can reflect the technological and economic relations among sectors of the national economy.

The three sections of the 1981 input–output (transactions) table include the intermediate quadrant, the final products or demand quadrant, and the primary inputs quadrant. The intermediate quadrant (24 by 24 sector) contains the industries producing goods and services in China. The role of this quadrant is to provide a representation of the economic interdependence among the producing sectors of China's economy. Since the input–output table is prepared in producers' prices, each entry in the intermediate quadrant represents an ex-factory value. The final products section of the input–output table is important because it represents the autonomous sector—the one in which changes occur which are transmitted throughout the rest of the table. Depreciation of fixed assets, personal income, and enterprise income are contained within the third major quadrant—the primary inputs quadrant—which lists inputs into each intermediate sector which originate outside the production system, that is, that are not purchased from firms within the local economy.

The national accounting and industrial classification scheme employed by the Chinese in compiling the 1981 input–output tables is the Material Balance System (MBS), rather than the System of National Accounts (SNA). The Soviets, too, traditionally used the MBS system in input–output analysis. According to Polenske (1986), the MBS limits the types of activities included to those that are 'productive', where only the purchases of 'productive' sectors are included in the intermediary portion of the accounts. Hence, all 'unproductive' sectors are excluded. Lengthy discussions of the MBS and SNA systems of accounting can be found in Polenske (1986), United Nations (1977*a*, *b*), and Stone (1970). A major difference between the two systems is the distinction between physical and non-physical production, which Western nations do not make.

According to the input–output table, the sum of purchases and sales or 'total gross output' in China for 1981 was 904.8bn. yuan (1981 $53.1bn.). Total gross output in China refers to total product of society or the sum total of gross output in value terms produced by five material-product sectors—agriculture, industry, construction, transport, and commerce (catering and supply and marketing of materials included). The value of the gross output of goods is recorded at the time the goods are produced by the sector; they may be sold or added to inventories in the form of either finished goods, or be part of the capital accumulation of that sector (Jensen and West 1986: 24).

In addition to total gross output, the input–output table lists the total final demand of China in 1981 at 435.2bn. yuan. It equals total value added (394.0bn. yuan) plus depreciation of fixed assets (41.1bn. yuan). Note that the entries in the intermediate sectors represent only those input requirements produced and used as inputs.

THE GRIMP COMPUTER PROGRAM: AN APPLICATION TO THE 1981 INPUT–OUTPUT TABLE OF CHINA

For this study, the GRIMP[4] computer program (West 1986; Jensen and West 1986) was applied to the 1981 input–output table of China in order to quantify the impact of China's minerals sector (specifically the metallurgical, coal and coke-making, and petroleum industries) on the nation's economy. The 24-sector input–output table is sufficiently disaggregated and detailed to illustrate via data manipulations the interdependence between mineral-related industries and other sectors of China's economy.[5] While it is clear that the specific interrelationships between the mineral-related industries and other sectors of the Chinese economy will change as economic and industrial policies evolve, analysis of the 1981 input–output table will reveal many significant fundamental relations that will not alter given the nature of a centrally planned economy. As a matter of consistency, much of the following discussion on structure, meaning, and solutions of China's input–output table uses terminology and definition of Jensen and West (1986).

Table 6.2 shows three columns of the larger direct coefficients table of China which, in complete form, contains 607 direct input coefficients. The coefficients matrix was calculated simply by dividing all of the entries in each sector's column

TABLE 6.2. *Excerpts of direct input coefficient matrix, China input–output table, 1981*

Sectors	Sector 6	Sector 8	Sector 9
1. Farming	0.000114	—	0.0009961
2. Forestry	0.000494	0.01383	0.00049
3. Animal husbandry	—	—	—
4. Sideline production	0.021259	0.020621	0.007836
5. Fishery	—	—	—
6. Metallurgical industry	0.320375	0.035302	0.023868
7. Power industry	0.073475	0.069962	0.022001
8. Coal and coke-making	0.049089	0.044206	0.000123
9. Petroleum industry	0.007532	0.00795	0.190693
10. Heavy chemical industry	0.007443	0.01738	0.030653
11. Light chemical industry	0.00204	0.004355	0.00218
12. Heavy machine-building industry	0.051905	0.072557	0.028483
13. Light machine-building industry	0.002665	0.006774	0.00623
14. Building materials industry	0.013962	0.015177	0.009631
15. Logging	0.000828	0.030848	0.0011
16. Manufacturing of wood and wood products	0.000055	0.000076	0.00052
17. Food	0.001301	0.001791	0.000397
18. Textile	0.001541	0.004485	0.003051
19. Clothing and leather	0.002603	0.007663	0.001896
20. Paper, cultural and educational articles	0.000974	0.001566	0.001186
21. Other manufacturing	0.016967	0.021414	0.00244
22. Construction	—	—	—
23. Transport, post, and telecommunications	0.043587	0.015465	0.021738
24. Trade, catering, supply, marketing of materials	0.0220075	0.012174	0.060439
25. Subtotal (1–24)	0.638283	0.403597	0.415952
26. Depreciation of fixed assets	0.066039	0.144696	0.102408
27. Personal income	0.070032	0.299124	0.028145
28. Enterprise income	0.225646	0.525831	0.453495
29. Subtotal (27 + 28)	0.295678	0.451706	0.48164
30. Total input (25 + 26 + 29)	1	1	1

Note: Rounding errors occur.

by the gross output for that industry. Each coefficient refers to the amount of input required from each of the 24 sectors at the left of the table by the three mineral-related industries listed across the top to produce a yuan of output. For example, each (or average) yuan of output of the metallurgical industry requires, from regional sources:

0.320 yuan in inputs from metallurgical industries,
0.073 yuan in inputs from the power industry,
0.052 yuan in inputs from heavy machine-building industries,

0.193 yuan in inputs from all other industries,
or a total of 0.638 yuan from all regional industries.

In addition:

0.66 yuan in depreciation of fixed assets,
0.070 yuan in personal income, and
0.226 yuan in enterprise income.

These are the direct or first-round purchases of inputs which do not include subsequent indirect effects on the regional economy. The direct coefficients corresponding to the intermediate quadrant—often referred to as the A matrix—represents the interdependence between the sectors of the local economy. Each aij coefficient in the A matrix therefore represents the amount of input required from sector i per unit (yuan) of output j (Jensen and West 1986: 42).

Using the direct input coefficients table of China, a calculation can be made of the amount of direct purchases required from each of the 24 industries along the left-hand side of the table as a result of an increase (or decrease) in the output of one or more of the industries at the top of the table. If, for example, the output of the coal and coke-making industry were increased by 1m. yuan (assuming constant technical coefficients), the direct inputs into the coal and coke-making industry (purchases from other industries) would be increased by the following amounts:

Inputs from industry	*Would be increased by:*
Heavy machine-building	72,557 yuan
Power industry	69,962 yuan
Coal and coke-making	44,206 yuan
Other industries	21,687 yuan

Notice that with a 1m. yuan increase in the output of the coal and coke-making industry, the total increase in the value of direct inputs into the same industry amounts to 208,412 yuan. If direct or technical coefficients remain relatively stable (or if they can be adjusted based on new data), a firm within an industry can determine how much it would have to buy directly from each of its supplying industries in order to boost production.

Table 6.2 shows the direct purchases that will be made by an industry from all other industries within the intermediate quadrant for each yuan's worth of current output. In addition to these direct or first-round purchases, an increase in the final demand for the products of an industry will generate both direct and indirect increases in the output of other industries in the intermediate quadrant. These direct and subsequent indirect increases eventually make their way through the economy, where each succeeding second, third, and subsequent-round effects become smaller and of less significance. Table 6.3 lists the direct and indirect effects of changes in the output of each sector, that is, the total effect of a yuan change in

TABLE 6.3. *Direct and indirect purchases, open general solution or open inverse matrix: China input–output table, 1981* (yuan)

Sector	Metallurgical industry	Coal and coke-making industry	Petroleum industry	Others
Metallurgical	1.52	0.10	0.07	1.68
Coal and coke-making	0.10	1.07	0.02	0.78
Petroleum	0.05	0.04	1.25	1.12
All others	0.76	0.57	0.44	38.63
TOTAL[a]	2.34	1.79	1.79	42.21

[a] Totals may not add owing to independent rounding.

the delivery to final demand of the output of each sector. There are several methods used for calculating the combined direct and indirect effects; an iterative procedure is commonly used.

In Table 6.3 it is seen that each (additional) yuan of sales by the metallurgical industry to final demand would result—subsequent to the many rounds of economic reaction throughout the economy—in 1.52 yuan being generated (including the assumed 1 yuan change) in the same sector, 0.10 (additional) yuan in the coal and coke-making industry, 0.05 (additional) yuan in petroleum production, and 0.76 (additional) yuan from all other sectors. This sums to a total of 2.34 yuan on the productive sectors of the economy as a whole. Each entry in Table 6.3 is therefore a linkage coefficient indicating an expected response in the economy to a yuan sales stimulus; the elements of the table are disaggregated output linkage coefficients showing the relative output effects which can be expected in all other sectors of the economy, and the sum of these items represents direct and indirect linkage coefficients, expressing the expected impacts on the entire economy.

The open general solution or open inverse lists the indirect or industrial support effects only, that is, increases attributable to production effects. The designation 'open' implies that the model includes only the intermediate sectors as previously defined, and none of the final demand sectors have been included in the A matrices (Jensen and West 1986: 51). Such a model depicts the situation where only the productive sectors of the economy are assumed to be endogenous to the system,[6] that is, when all final demand sectors are assumed to be determined by factors outside the regional productive system (Jensen and West 1986: 53). If this assumption is unsatisfactory, the model can be fully or partially closed with respect to other factors—for example, to include consumption-induced effects. In this analysis, the input–output table is closed with respect to households by bringing the personal income sector into the intermediate quadrant. By wholly incorporating the personal income sector into the closed model, it is assumed that essentially all of the income earned locally is spent locally. Personal income in the input–output table of China represents return to household labour in the form of wages and salaries (income), and relatively little from a government welfare fund for staff workers.

The elements of the closed inverse matrix (Table 6.4) are, as with the open inverse matrix, disaggregated output linkage coefficients. As seen, these numbers are larger than those in Table 6.3 because they include output levels required by industries to meet the consumption-induced output effects which came about by closing the input–output model with respect to households. Table 6.4 can be interpreted using the petroleum sector as an example. For each yuan of petroleum output, direct, indirect (industrial support), and consumption-induced effects add to 0.08 yuan output from the metallurgical industry, 0.02 yuan from coal and coke-making, and so on, or an output effect over all productive sectors of 2.27 yuan, including the original yuan.

The household (personal) income row generated by GRIMP shows the amount of income which accrues to households in direct, indirect, and consumption-induced effects, per yuan of sales by each sector. As may be expected, more income is generated to the household (personal) sector per yuan of sales of the coal and coke-making industry than the metallurgical and petroleum industries combined. This also implies that the labour inputs necessary to produce a unit of coal or coke are more costly than those required to produce a unit of output from the metallurgical or petroleum industries. Altogether, the combined direct, indirect, and consumption-induced effects (total output effects) of the coal and coke-making industry (3.51) are greater than those generated by the metallurgical industry, and nearly 35 per cent greater than that of the petroleum industry, illustrating the strongly felt effects of coal output on the Chinese economy. The household (personal) consumption in Table 6.4 indicates a consumption multiplier, that is, the effect on the output of each sector of an average or additional yuan of household consumption in the region (Jensen and West 1986: 55).

TABLE 6.4. *Direct, indirect, and consumption-induced purchases, closed general solution or closed inverse matrix: China input–output table, 1981* (yuan)

Sector	Metallurgical industry	Coal and coke-making industry	Petroleum industry	Others	Household consumption
Metallurgical	1.55	0.14	0.08	2.61	0.12
Coal and coke-making	0.12	1.10	0.02	1.34	0.06
Petroleum	0.06	0.07	1.26	1.73	0.07
All others (subtotal)[a]	(3.32)	(3.51)	(2.27)	(77.08)	(4.1)
Household Income[b]	0.51	0.89	0.25	18.04	2.13

[a] Subtotals may not add owing to independent rounding.

[b] Designated as personal income (row 27) in the input–output table (see Table 6.2 above).

OUTPUT AND INCOME LINKAGE COEFFICIENTS

The subtotals presented in Table 6.4 are known as total output linkage coefficients; here total indicates that the initial, direct, indirect, and consumption-induced effects of an increase in output of one sector are included. The linkage coefficients discussed in the foregoing paragraphs are sometimes referred to as 'multipliers' in input–output literature. However, because the linkage coefficients do not represent traditional Keynesian-type multipliers (spreading of an economic shock resulting from an injection of income), they are not referred to as multipliers in this study in order to distinguish between the two types.[7]

Two sets of linkage coefficients were computed in this study—output and income. Table 6.5 lists the output and income linkage coefficients for the metallurgical industry (sector 6), coal and coke-making industry (sector 8), and the petroleum industry (sector 9). The output linkage coefficients indicate how much

TABLE 6.5. *Output and income linkage coefficients,[a] China input–output table, 1981*

	Initial effect	First round	Industrial support	Consumption induced	Total linkage coefficient	Flow-on
A. Output linkage coefficients[b]						
6. Metallurgical	1.00	0.64	0.70	0.98	3.32	2.32
8. Coal and coke-making	1.00	0.40	0.39	1.72	3.51	2.51
9. Petroleum	1.00	0.42	0.37	0.48	2.27	1.27
B. Income linkage coefficient[b]						
6. Metallurgical	0.07	0.07	0.10	0.27	0.51	0.44
8. Coal and coke-making	0.30	0.06	0.06	0.47	0.89	0.59
9. Petroleum	0.03	0.04	0.05	0.13	0.25	0.22
C. Ratios (income)	Type IA[c]		Type IB[d]	Type IIA[e] (total)		Type IIB[f] (flow-on)
6. Metallurgical	2.02		3.41	7.26		6.26
8. Coal and coke-making	1.20		1.40	2.98		1.98
9. Petroleum	2.33		4.17	8.88		7.88

Notes: Type I and Type II ratios are computed to determine relationships between inital income (own sector) and flow-on effects.

Type I and Type II ratios are not necessary for output multipliers considering they have a unity base.

[a] Also referred to as 'multipliers'.
[b] Output per yuan of output of each sector.
[c] Type IA = (Initial effect + First-round effect)/(Initial effect)
[d] Type IB = (Initial effect + First-round effect + Industrial support)/(Initial effect)
[e] Type IIA = Total effect/Initial effect
[f] Type IIB = Flow-on effect/Initial effect

total production in the economy will increase for every 1 yuan increase in production of each sector; that is, the production corresponding to total output requirements associated with a 1 per cent increase in final demand. These linkage coefficients show the sectoral impact of a stimulus, and allow for estimates of the expected reaction on each and every sector in the Chinese economy to a proposed change in the final demand of one sector in the economy. Income linkage coefficients reveal how much income is generated in an economy given a change in final demand; that is, the total income in value added associated with a 1 per cent increase in final demand. An output linkage coefficient represents a column total of the Leontief inverse matrix, whereas an income coefficient is a vector of value added premultiplied by the same Leontief inverse matrix.

The 24-sector input–output table of China does not contain employment figures by sector, making it difficult to calculate employment linkage coefficients. If desired, however, such coefficients could be determined by utilizing China's 1982 census statistics, or by using value of labour data (wage compensation coefficients). However, since it is admittedly difficult to estimate accurately the labour forces of the three mineral-related industries (particularly the coal and coke-making sectors), it was decided not to generate employment linkage coefficients in this study. There are virtually thousands of collectively and individually run coal- mines in China which employ millions of workers at various tasks, and it is difficult to estimate the number of workers at an accuracy required for input–output analysis.

In addition, many of the activities related to mining in China are essentially viewed as social programmes, where full or near-full employment is a primary objective. As a consequence, a change in the output of a particular sector does not necessarily imply a proportional change in labour usage, at least in the short term. Therefore, employment linkage coefficients may not properly indicate the labour responses given a change in output.

In an in-depth article on China's labour-force development since 1952, Taylor (1986) documented the various means of employment creation in China to alleviate unemployment problems. These have included the permission of overstaffing in state enterprises (known as 'feather-bedding' labour practices) to accommodate periodic gluts of new entrants to the job market. During such times labour productivity in state-owned industrial enterprises often grew more slowly than expected, indicating that workers were being hired beyond the point of positive contributions to output.[8] Thus an overall slowing of industrial output growth could have occurred along with an increase in labour use.

Column 6 in Table 6.5 lists the flow-on effects of the output and income linkage coefficients, which represent the differences between the total coefficients and their initial effects. In essence, a flow-on effect is a measurement of the response to an initial effect or stimulus, that is, the output or (household) income result of an assumed 1 yuan output stimulus to an economy. The flow-on effects of the output linkage coefficients range between 1.27 for the petroleum industry and 2.51 for coal and coke-making, indicating that, when an increase in output from the petroleum industry occurs, the output result of the stimulus to the Chinese economy is lower than that which would occur with an output stimulus from the other two sectors. The flow-on effects include consumption-induced effects which in many cases can be significant. As seen in Table 6.5, the consumption-induced output effect of the coal and coke-making industry (1.72) is significantly greater than that of the other two mineral-related industries. The relatively low flow-on output effect of the petroleum sector is due to the highly specialized equipment and operating materials that are used in the industry, some of which must be imported. In contrast, the metallurgical and coal and coke-making industries exhibit a greater degree of interdependence within the Chinese economy and require more input from a greater number of sectors.

In addition to output linkage coefficients, income coefficients for the three mineral-related industries are also shown in Table 6.5. These reveal the income effects (or changes in value added) of an output change; that is, that different amounts of income are generated by different sectors of the economy even if we assume that each sector expands its output by the same amount. As indicated previously, the total income effect of an assumed yuan increase of output of a sector is provided in the last row of the closed inverse matrix (Table 6.4). For example, as indicated in the bottom row of Table 6.4, each yuan of metallurgical sales to final demand contributes 0.51 yuan to household (personal) income. Similarly, sales from the coal and coke-making industry yield 0.89 yuan to household income, while petroleum sales yield much less—0.25 yuan. The greater the extent of interdependence within the Chinese economy or, conversely, the less the dependence on imports, the greater will be the direct income changes.

Table 6.5 disaggregates the income linkage coefficients by initial, direct, indirect, consumption-induced, and flow-on effects. According to the table, each yuan of metallurgical industry sales to final demand results in 0.07 yuan in income within the same sector, 0.07 yuan in industries affected by direct (first-round) purchases, 0.10 yuan in industries affected by indirect (industrial support) effects, 0.27 yuan in industries affected by consumption-induced effects, giving a total income effect of 0.51 yuan per yuan of metallurgical industry output. Of this, 0.44 yuan occurs as a result of flow-on output effects in the many sectors of the economy.

The lower part of Table 6.5 presents Type I and Type II income ratios—in order to illustrate the relationship between initial or own-sector income effects and flow-on income effects. Type I and Type II direct and indirect ratios can be defined as follows:

$$\text{Type I Ratio} = \frac{(\text{Initial} + \text{Final Round} + \text{Industrial Support})}{\text{Initial (own-sector) Effect}}$$

$$\text{Type IIA Ratio} = \frac{\text{Total Effect}}{\text{Initial Effect}}$$

$$\text{Type IIB Ratio} = \frac{\text{Flow-on Effects}}{\text{Initial Effects}}$$

Note that there is not a direct line of causation between Type I and Type II ratios; that is, the own-sector household income component does not cause the flow-on component (Jensen and West 1986: 61). Both are the result of the assumed yuan of output change. According to Table 6.5, each yuan of household income accruing to labourers in the coal and coke-making industry, for example, is associated with 1.98 yuan in household income in those sectors which have flow-on output effects from the coal and coke-making industry. This association between sectors leads some input–output analysts to refer to Type I and Type II ratios as 'measures of association' (Jensen and West 1986: 62).

Although Table 6.5 lists the flow-on output and income effects of the three mineral-related industries in China, it does not disaggregate those effects by the 24 sectors modelled in the input–output table. However, the GRIMP computer program does assemble and rank in descending order the flow-on effects resulting from an output stimulus in any sector. Table 6.6 shows the ranked output and

TABLE 6.6. *Ranked output and income flow-ons, stimulus in coal and coke-making industry: China input–output table, 1981* (yuan)

Rank	Sector[a]	%	Flow-on output linkage coefficient	Sector[a]	%	Flow-on income linkage coefficient
1.	Farm	20.34	0.51	Farm	50.89	0.30
2.	Textile	8.70	0.22	AnmHub	11.80	0.07
3.	Food	8.60	0.22	TrCaMark	5.00	0.03
4.	HeaMBui	6.03	0.15	CoalCoke	4.89	0.03
5.	MetalInd	5.73	0.14	Forest	3.65	0.02
6.	HeaChem	5.41	0.14	SideProd	3.27	0.02
7.	TrCaMark	5.26	0.13	HeaMBui	3.00	0.02
8.	PowInd	4.99	0.13	TranspP	2.39	0.01
9.	AnmHub	4.81	0.12	MetalInd	1.70	0.01
10.	CoalCok	3.86	0.10	Textile	1.56	0.01
11.	SideProd	3.68	0.09	Logging	1.41	0.01
12.	LigMBui	3.11	0.08	LigMBui	1.25	0.01
13.	Petro	2.61	0.07	HeaVChem	1.25	0.01
14.	TranspP	2.43	0.06	Food	1.19	0.01
15.	LigChem	2.37	0.06	OtherM	1.16	0.01
16.	OtherM	2.30	0.06	Fish	1.00	0.01
17.	ClothLe	2.03	0.05	BuilMat	0.91	0.01
18.	Logging	1.98	0.05	PowInd	0.77	0.00
19.	Forest	1.56	0.04	ClothLe	0.67	0.00
20.	PpCulEd	1.41	0.04	PpCulEd	0.67	0.00
21.	BuilMat	1.20	0.03	Constr	0.59	0.00
22.	Constr	0.67	0.02	LigChem	0.40	0.00
23.	Fish	0.56	0.01	Petro	0.31	0.00
24.	ManuWood	0.37	0.01	ManuWood	0.27	0.00
TOTAL		100.0	2.51	TOTAL	100.0	0.59

[a] For complete sector name see Table 6.2 above.

income flow-ons assuming a stimulus generated within the coal and coke-making industry. This table shows the sectoral impacts of an assumed 1 yuan output stimulus in the industry in terms of percentage of reaction and flow-on linkage coefficients by sector.

In terms of output, Table 6.6 shows that after an output stimulus in the coal and coke-making industry, 20.34 per cent of the flow-on output effects occur in the farming sector, 8.70 per cent in the textile sector, and so on. With respect to income, a remarkable 50.89 per cent of the flow-on income effects occur in the farming sector. One reason why the farming sector may feel such a strong impact (backward linkage effect) when an increase in coal output occurs is because before 1985, all income received by farmers was designated as farming income which, according to China's State Statistical Bureau, was part of a broader category of agricultural income. By definition, total agricultural was disaggregated into farming, forestry, animal husbandry, sideline occupation, and fishery income, and also included disposable income from other productive and non-productive sideline activities—such as coal-mining. Note that after 1985, all income generated by coal-mining—even by farmers—is categorized as earnings in the coal and coke-making industry.

Farmers earn income from coal-mining in one of two ways: (1) by receiving compensation from other workers mining on their own farmland, or (2) by mining coal individually or collectively in a group. In either case, as of 1979, the state purchasing price of coal almost doubled, which meant that farmers received more income per unit of coal mined. This income, in turn, had secondary effects which generated even more income to the farmers. More capital was made available to the farmers to reinvest in mechanization of agricultural processes, expansion of coal-mining activities, or other rural activities—with the returns of all of these activities being counted as agricultural income until 1986. This may partially explain why 50.89 per cent of the flow-on income effects of an increase in coal output occurs in the farming sector.

CONCLUDING REMARKS ABOUT LINKAGE COEFFICIENTS

Each sector within the Chinese economy has the potential to generate new income and output (and employment) in the nation given a change in sales to final demand or domestic sectoral output. If, for example, sales to final demand (exports) increase, income may be generated throughout the economy (backward linkage effect). Or, if a rise in domestic sectoral output occurs, this additional output may serve as inputs to other industries and undergo further processing and (forward linkage effect). Policymakers, however, cannot simply select the sector with the largest linkage coefficient as the candidate to expand an economy. There must be sufficient demand for the goods from any sector being targeted for increased output in order to expand the economy. Since China's economy in recent years has been growing at a relatively rapid pace, sufficient demand for increased output of a particular sector could feasibly be generated from within. However, some sectors—such as the metallurgical industry—have export potential for certain speciality metals, such as the rare earths, and therefore should be subject to further development. If the metallurgical industry, for example, could export more semi-finished to finished goods, much of the value added of the processed goods could remain within the domestic economy.

IMPACTS OF A CHANGE IN ONE OR MORE ELEMENTS OF FINAL DEMAND

In the foregoing discussion the flow-on output or income effects analysed were in response to a proposed change in output of one particular sector of the Chinese economy. Often these changes occur following a change in one of the final demand sectors, such as consumption, capital accumulation, or exports. As an example, a change in the level of exports will stimulate the exports sector initially, which is then felt by the producing sectors of an economy as they respond to the many production impulses passed through the economy. Some input–output analysts (Jensen and West 1986: 97) believe that most policy problems can be reduced to computing the effect of a change in one or more elements of final demand on the producing sectors of an economy. Note that in traditional input–output modelling, a distinction does not have to be made between the various final demand categories. For instance, a 1 yuan increase in exports (column 33) will have the same effect as a 1 yuan increase in personal consumption (column 27).

The inverted Leontief matrix (also known as the matrix of direct and indirect requirements per yuan of final demand when transposed) can be used to estimate the total impact on an economy of changes in one or more elements of final demand. The Leontief matrix is calculated from a series of simultaneous equations written in matrix format (Jensen and West 1986: 98–104):

(1) $$AX + Y = X$$

where A = {aij} is the matrix of direct coefficients,
X = the column vector of sector gross output (column totals), and
Y = the column vector of total final demands.

Subtracting the term AX from both sides of the equation yield:

(2) $$Y = X - AX$$

(3) $$Y = (I - A)X$$

(4) or $$X = (I - A) - 1Y$$

where (I − A) − 1 is the general solution or Leontief inverse matrix.

Solving this system of equations allows the impact of a change in one element of final demand to be estimated. For numerous changes in final demand, equation (4) must be expanded in terms of the direct, indirect, and consumption-induced (closed) output effects.

For this study, the aggregated and disaggregated impacts of a change in one element of final demand, namely exports, were determined for the three mineral-related industries of China (Tables 6.7, 6.8, and 6.9). In Table 6.7, it was assumed that exports from the metallurgical industry of China declined by 180m. yuan. This was, according to China's Customs Statistics, the estimated decrease in export value from the industry between 1981—the year of the input–output table—and 1982. Export values are, however, determined using world prices whereas transactions represented in the input–output table are expressed using transaction prices. None the less, the percentage decline of metallurgical industry exports from 1981 to 1982 used here provides a realistic assumption of a change in final demand sales to exports. Note that technological changes and changes in purchasing patterns among China's producing sectors between 1981 and 1982 are not accounted for in the output and income impacts shown in Table 6.7.

As seen in Table 6.7, the resultant aggregated impact is the change in sector output and income. For example, the initial decrease of 180m. yuan in final demand sales results in a decrease of 115.2m. yuan in direct or first-round effects, 126m. yuan in indirect or industrial support effects, and 176.4m. yuan in consumption-induced effects, giving a total flow-on effect of negative 417.6m. yuan. This flow-on effect is distributed throughout the 24 producing sectors as listed in the lower part of Table 6.7. Notice that the decline of exports from the metallurgical industry of 180m. yuan results in a negative flow-on impact on the same sector of 99m. yuan; the remaining negative impacts are felt by other sectors of the economy in lesser amounts.

Table 6.7 also lists the income impacts which occur when one element of final demand changes, that is, a 180m. yuan decrease in export value from the metallurgical industry. The initial decline of income of 12.6m. yuan leads to further direct, indirect, and consumption-induced effects—which, combined, represent a decline of income of 91.8m. yuan within the Chinese economy.

The impacts of prospective increases in exports from the coal and coke-making and petroleum industries are summarized in Tables 6.8 and 6.9. As with the metallurgical industry, the assumed increases in final demand sales to exports for the coal and coke-making and petroleum industries represent realistic percentage changes in final demand sales. Table 6.8 shows the aggregated and disaggregated impacts of an increase in export value of coal of 134m. yuan; while Table 6.9 illustrates the impacts of an 834m. yuan increase in petroleum exports.

In reality, impact situations are more complex than a simple change in one element of final demand. Quite often multiple changes in final demand occur leading to a complex interaction of output and income responses throughout an economy. These adjustments may represent a general move in one direction for all final demand categories, such as an overall increase in consumption (personal and public) as a result of a major tax reduction.

The impacts on the Chinese economy of changes in final demand of two sectors—the coal and coke-making and petroleum industries—are presented in detail in Table 6.10. This example is provided to demonstrate how multiple

TABLE 6.7. *Aggregated and disaggregated impacts of a change[a] in one element of final demand: metallurgical industrial exports*

	Output impacts		Income impacts	
	Linkage coefficient (1)	Amount[b,c] (2)	Linkage coefficient (3)	Amount[b,d] (4)
A. Aggregated impacts[e]				
Initial effect	1.00	−180.00	0.07	−12.60
First round	0.64	−115.20	0.07	−12.60
Industrial support	0.70	−126.00	0.10	−18.00
Consumption induced	0.98	−176.40	0.27	−48.60
TOTAL	3.32	−597.60	0.51	−91.80
Flow-on	2.32	−417.60	0.44	−79.20
B. Disaggregated flow-on impacts[f]				
Farm	0.30	−54.00	0.18	−32.40
Forest	0.02	−3.60	0.01	−1.80
AnmHub	0.07	−12.60	0.04	−7.20
SideProd	0.08	−14.40	0.02	−3.60
Fish	0.01	−1.80	0.00	0.00
MetalInd	0.55	−99.00	0.04	−7.20
PowInd	0.15	−27.00	0.01	−1.80
CoalCok	0.12	−21.60	0.04	−7.20
Petro	0.06	−10.80	0.00	0.00
HeaChem	0.09	−16.20	0.00	0.00
LigChem	0.04	−7.20	0.00	0.00
HeaMBuil	0.14	−25.20	0.02	−3.60
LigMBuil	0.05	−9.00	0.00	0.00
BuilMat	0.03	−5.40	0.01	−1.80
Logging	0.01	−1.80	0.00	0.00
ManuWood	0.01	−1.80	0.00	0.00
Food	0.13	−23.40	0.00	0.00
Textile	0.13	−23.40	0.01	−1.80
ClothLe	0.03	−5.40	0.00	0.00
PpCulEd	0.02	−3.60	0.00	0.00
OtherM	0.05	−9.00	0.01	−1.80
Constr	0.01	−1.80	0.00	0.00
TranspP	0.10	−18.00	0.02	−3.60
TrCaMar	0.11	−19.00	0.02	−3.60
TOTAL	2.32	−417.60	0.44	−79.20

Notes: Totals may not add owing to independent rounding.
For complete sector names see Table 6.2.

[a] Assumption is that metallurgical industrial exports decreased by 180m. yuan to the next year.
[b] Amount in million yuan.
[c] −180m. yuan multiplied by column (1).
[d] −180m. yuan multiplied by column (3).
[e] Part A is from Table 6.5.
[f] Part B is from Table 6.6.

TABLE 6.8. *Aggregated and disaggregated impacts of a change*[a] *in one element of final demand: coal and coke-making industrial exports*

	Output impacts		Income impacts	
	Linkage coefficient (1)	Amount[b,c] (2)	Linkage coefficient (3)	Amount[b,d] (4)
A. Aggregated impacts[e]				
Initial effect	1.00	+134.00	0.30	+40.20
First round	0.40	+53.60	0.06	+8.04
Industrial support	0.25	+33.50	0.06	+8.04
Consumption induced	1.72	+230.48	0.47	+62.98
TOTAL	3.51	+470.34	0.89	+119.26
Flow-on	2.51	+336.34	0.59	+79.06
B. Disaggregated flow-on impacts[f]				
Farm	0.51	+68.34	0.30	+40.20
Forest	0.04	+5.36	0.02	+2.68
AnmHub	0.12	+16.08	0.70	+9.38
SideProd	0.09	+12.06	0.02	+2.68
Fish	0.01	+1.34	0.01	+1.34
MetalInd	0.14	+18.76	0.01	+1.34
PowInd	0.13	+17.42	0.00	0.00
CoalCok	0.10	+13.40	0.03	+4.02
Petro	0.07	+9.38	0.00	0.00
HeaChem	0.14	+18.76	0.01	+1.34
LigChem	0.06	+8.04	0.00	0.00
HeaMBuil	0.15	+20.10	0.02	+2.68
LigMBuil	0.08	+10.72	0.01	+1.34
BuilMat	0.03	+4.02	0.01	+1.34
Logging	0.05	+6.70	0.01	+1.34
ManuWood	0.01	+1.34	0.00	0.00
Food	0.22	+29.48	0.01	+1.34
Textile	0.22	+29.48	0.01	+1.34
ClothLe	0.05	+6.70	0.00	0.00
PpCulEd	0.04	+5.36	0.00	0.00
OtherM	0.06	+8.04	0.01	+1.34
Constr	0.02	+2.68	0.00	0.00
TranspP	0.06	+8.04	0.01	+1.34
TrCaMar	0.13	+17.42	0.03	+4.02
TOTAL	2.51	+336.34	0.59	+79.06

Notes: Totals may not add owing to independent rounding.
For complete sector names see Table 6.2.

[a] Assumption is that exports of the coal and coke-making industry increased by 134m. yuan to the next year.
[b] Amount in million yuan.
[c] 134m. yuan multiplied by column (1).
[d] 134m. yuan multiplied by column (3).
[e] Part A is from Table 6.5.
[f] Part B is from Table 6.6.

TABLE 6.9. *Aggregated and disaggregated impacts of a change*[a] *in one element of final demand: petroleum industrial exports*

	Output impacts		Income impacts	
	Linkage coefficient (1)	Amount[b,c] (2)	Linkage coefficient (3)	Amount[b,d] (4)
A. Aggregated impacts[e]				
Initial effect	1.00	+824.00	0.03	+24.72
First round	0.42	+386.08	0.04	+32.96
Industrial support	0.37	+304.88	0.05	+41.20
Consumption induced	0.48	+395.52	0.13	+107.12
TOTAL	2.27	+1,870.48	0.25	+206.00
Flow-on	1.27	+1,046.48	0.22	+181.28
B. Disaggregated flow-on impacts[f]				
Farm	0.16	+131.84	0.09	+74.16
Forest	0.01	+8.24	0.01	+8.24
AnmHub	0.04	+32.96	0.02	+16.48
SideProd	0.04	+32.96	0.01	+8.24
Fish	0.00	0.00	0.00	0.00
MetalInd	0.08	+65.92	0.01	+8.24
PowInd	0.05	+41.20	0.00	0.00
CoalCok	0.02	+16.48	0.01	+8.24
Petro	0.26	+214.24	0.01	+8.24
HeaChem	0.09	+74.16	0.00	0.00
LigChem	0.02	+16.48	0.00	0.00
HeaMBuil	0.07	+57.68	0.01	+8.24
LigMBuil	0.03	+24.72	0.00	0.00
BuilMat	0.02	+16.48	0.00	0.00
Logging	0.01	+8.24	0.00	0.00
ManuWood	0.01	+8.24	0.00	0.00
Food	0.07	+57.68	0.00	0.00
Textile	0.07	+57.68	0.00	0.00
ClothLe	0.02	+16.48	0.00	0.00
PpCulEd	0.02	+16.48	0.00	0.00
OtherM	0.00	+16.48	0.00	0.00
Constr	0.00	0.00	0.00	0.00
TranspP	0.05	+41.20	0.01	+8.24
TrCaMar	0.12	+98.88	0.03	+24.72
TOTAL	1.27	+1,046.48	0.22	+181.28

Notes: Totals may not add owing to independent rounding.
For complete sector names see Table 6.2.

[a] Assumption is that exports of the petroleum industry increased by 824 m. yuan to the next year.
[b] Amount in million yuan.
[c] 824m. yuan multiplied by column (1).
[d] 824m. yuan multiplied by column (3).
[e] Part A is from Table 6.5.
[f] Part B is from Table 6.6.

changes in final demand can, at times, offset one another or, at other times, complement each other in terms of the nature of the responses in the economy. As an example, as with Table 6.8 assume that additional coal and coke-making industry exports amounting to 134m. yuan are to take effect in a following year (1982 in this case). Also assume that at the same time there will be a decrease in the exports of petroleum industrial output by 134m. yuan. Note that in actuality it is unlikely that the increases and decreases of two elements of final demand will be equally offsetting; it is assumed here for convenience. None the less, the example is realistic considering that the export value of coal and coke in China has risen throughout the 1980s and, with a recent decline in the price of oil and quantity of exports, revenues from petroleum exports in China have dropped.

In a very simple economic model, because the net change of export value in this example is zero, it might be concluded that the net effects felt throughout the economy would likewise be minimal. However, because (in terms of output and income) the coal and coke-making industry affects the other sectors of the Chinese economy differently than the petroleum industry, the net economic changes are not zero. According to Table 6.10, the flow-on output effects by the coal and coke-making industry is more than 336m. yuan. This greatly exceeds the decline in flow-on output (170.2m. yuan) brought about by the petroleum industry. The total net increase in the Chinese economy is 166.2m. yuan. Such an increase occurs because, as stated previously, the coal and coke-making industry—although labour-intensive—is more interrelated with other producing sectors of the Chinese economy than the petroleum industry. This was shown by comparing the input coefficients of the two industries (Table 6.2).

Like the flow-on output impacts of the two elements change in final demand, the flow-on income effects are positive. As seen in Table 6.10, there is an initial increase in income of 36.2m. yuan, which eventually translates into a total income increase of 85.8m. yuan generated throughout the economy. This increase in income represents a redistribution of income from the petroleum industry to the coal and coke-making industry as a result of the changes in export patterns. Additionally, the redistribution of income is compounded by the higher relative (personal) income rates of coal-miners versus workers in the petroleum industry. According to Table 6.2, the input coefficient to personal income (row 27) for the coal and coke-making industry is 0.299, compared with 0.028 for the petroleum industry.

Considering the net economic change of an increase in coal exports and a decrease in petroleum is overwhelmingly positive in terms of flow-on output and income effects, the current policy of the Chinese government to encourage exports of crude oil and refined petroleum products is perhaps misguided. The input–output results just discussed suggest that it may be more beneficial to the overall national economy to encourage additional coal production and, in turn, exports abroad. Before considering this however, the level of demand for increased coal production and exports must be assessed, with strong consideration of the quality of coal available in China. In addition, there are good reasons to continue

TABLE 6.10. *Aggregated and disaggregated impacts of a change*[a] *in two elements of final demand: coal and coke-making and petroleum industrial exports*

	Output impacts[b]			Income impacts[b]		
	Coal ind. (1)[c]	Petro. ind. (2)[d]	Net change (3)[e]	Coal ind. (4)[c]	Petro. ind. (5)[d]	Net change (6)[f]
A. Aggregated impacts						
Initial effect	+134.00	−134.00	0.00	+40.20	−4.02	+36.18
First round	+53.60	−56.28	−2.68	+8.04	−5.36	+2.68
Industrial support	+33.50	−49.58	−16.08	+8.04	−6.70	+1.34
Consumption induced	+230.48	−64.32	+166.16	+62.98	−17.42	+45.56
TOTAL	+470.34	−304.18	+166.16	+119.26	−33.50	+85.76
Flow-on	+336.34	−170.18	+166.16	+79.06	−29.48	+49.58
B. Disaggregated flow-on impacts						
Farm	+68.34	−21.44	+46.90	+40.20	−12.06	+28.14
Forest	+5.36	−1.34	+4.02	+2.68	−1.34	+1.34
AnmHub	+16.08	−5.36	+10.72	+9.38	−2.68	+6.70
SidePro	+12.06	−5.36	+6.70	+2.68	−1.34	+1.34
Fish	+1.34	0.00	+1.34	+1.34	0.00	0.00
MetalInd	+18.76	−10.72	+8.04	+1.34	−1.34	0.00
PowInd	+17.42	−6.70	+10.72	0.00	0.00	0.00
CoalCok	+13.40	−2.68	+10.72	+4.02	−1.34	+2.68
Petro	+9.38	−34.84	−25.46	0.00	−1.34	−1.34
HeaChem	+18.76	−12.06	+6.70	+1.34	0.00	+1.34

LigChem	+8.04	−2.68	+5.36	0.00	0.00	0.00
HeaMBuil	+20.10	−9.38	+10.72	+2.68	−1.34	+1.34
LigMBuil	+10.72	−4.02	+6.70	+1.34	0.00	+1.34
BuilMat	+4.02	−2.68	+1.38	+1.34	0.00	+1.34
Logging	+6.70	−1.34	+5.36	+1.34	0.00	+1.34
ManuWood	+1.34	−1.34	0.00	0.00	0.00	0.00
Food	+29.48	−9.38	+20.10	+1.34	0.00	+1.34
Textile	+29.48	−9.38	+20.10	+1.34	0.00	+1.34
ClothLe	+6.70	−2.68	+4.02	0.00	0.00	0.00
PpCulEd	+5.36	−2.68	+2.68	0.00	0.00	0.00
OtherM	+8.04	−2.68	+5.36	+1.34	0.00	+1.34
Constr	+2.68	0.00	+2.68	0.00	0.00	0.00
TranspP	+8.04	−6.70	+1.34	+1.34	−1.34	0.00
TrCaMar	+17.42	−16.08	+1.34	+4.02	−4.02	0.00
TOTAL	+336.34	−170.18	+166.16	+79.06	−29.48	+49.58

Notes: Total may not add owing to independent rounding.
For complete sector names see Table 6.2.

[a] Assumption is that from one year to the next exports of the coal and coke-making industry increased by 134m. yuan while concurrently petroleum exports decreased by the same amount.

[b] Amount in million yuan.

[c] +134m. yuan multiplied by coal industry linkage coefficient from Table 6.5.

[d] −134m. yuan multiplied by petroleum industry linkage coefficient from Table 6.5.

[e] Column (1) plus column (2).

[f] Column (4) plus column (5).

exporting petroleum products, including the relatively rapid generation of foreign exchange, and the tremendous value added received by refining crude oil and selling products. None the less, the question of which exports can be more beneficial to China's overall economy warrants further study.

OVERVIEW OF MAIN FINDINGS

The input–output modelling discussed in this chapter was done to illustrate the economic impacts on the Chinese economy of the metallurgical, petroleum, and coal industries. Combined, these three industries traditionally represent a significant portion of China's annual GVIO or value of industrial output. The modelling is considered preliminary in nature, as the application of input–output analysis in China is still in its early stages, and there is a general lack of a comprehensive set of economic statistics in the country.[9] The modelling described was based on several assumptions, including the expectation that Chinese industry typically strives for expanded production or output. It is additionally assumed that the actual transactions prices represented in the input–output tables reflect scarcity of resources. However, given the current pricing system in China, prices do not necessarily reflect real costs to the economy of resource usage. If different prices were indeed used in the modelling procedure, the resulting income and output linkage coefficients would be different.

Based on the input–output analysis, the coal and coke-making industry is more interrelated with the other 23 sectors of China's economy than either the metallurgical or the petroleum industries. According to the direct coefficients matrix presented in Table 6.2, purchases of inputs by the coal and coke-making industry are greater than those of the metallurgical and petroleum industries for 11 of the 24 sectors in the China input–output table. In contrast, both the metallurgical and petroleum industries purchase more inputs than the other two mineral-related industries in only five sectors.

In terms of income generation, the coal and coke-making industry boasts higher relative income rates than the metallurgical and petroleum industries, as indicated by the coefficients in row 27 of Table 6.2. Each yuan of output of the coal and coke-making industry requires 0.30 yuan in salaries and wages, as compared to 0.07 and 0.03 yuan for the metallurgical and petroleum industries respectively.

Even though the coal and coke-making industry appears to be more interrelated with other sectors of the Chinese economy than either the metallurgical or the petroleum industry, its overall impact on the nation's economy is not as great as that of many agricultural sectors, particularly farming. In terms of direct impact, the coal and coke-making industry contributed 0.49 per cent of the total output in 1981 as defined in the input–output table, whereas the farming sector contributed much more—32.7 per cent. With respect to household income, the coal industry contributed 2.27 per cent of the total compared to more than 45 per cent by the farming sector. A more comprehensive study of the direct contribution of a sector would include an evaluation of the sector's entries in the transactions table, as well as an examination of the A matrix for a better understanding of purchasing patterns. It should also be noted that any comprehensive analysis of the economic impact of a sector or sectors on the Chinese economy would include not only the economic significance of primary linkages in the economy, but also the indirect effects attributable to the industry or sector in question (Jensen and West 1986: 84).

Because the input–output results presented in this chapter provide only a 'snapshot' perspective of the relationship between mining and the Chinese economy, an in-depth historical analysis of the role of minerals in the economic development of China was provided in Chapter 2. With the introduction of newer and expanded input–output tables of China, more detailed analyses of the mineral-related industries of China and their contribution to the nation's economy will soon become possible. Such analyses can further examine the backward and forward linkages of the metallurgical, petroleum, and coal industries, and the extent to which increased government investment in these industries might be warranted.

INPUT–OUTPUT ANALYSIS AND OTHER TECHNIQUES

A commonly cited flaw of input–output analysis is that it presents a mere 'snapshot' view of an economy in time. Given changes in technology, input prices, and so on, an input–output table of one year should be adjusted based on new information to predict, for example, future input requirements. Technological changes need to be accounted for, as well as changes in purchasing patterns of industries. Additionally, prices of inputs may change relative to one another, resulting in changes in production processes.

Despite these criticisms, however, used in conjunction with modern computer technology, inter-industry analysis can be a very powerful and analytical tool. But it must be used carefully with a full understanding of its weaknesses and problems. Input–output research continues today in China, where interest was rejuvenated during the 1980s with the advent of new economic policies. In China, input–output models and other mathematical techniques are currently viewed as a way to help assess and plan for the economic transformation of the country (Polenske 1986). Recent efforts to reform the price system of China have spawned input–output studies by the State Council and State Price Bureau to model the economic impacts of reform and make new price reforms set for the 1990s more realistic. Effective utilization of input–output tables can help rectify relative prices among sectors in particular.

Modelling an economy in an effort to characterize the minerals industry is one application of input–output analysis. The technique allows the quantification of all the effects on an economy of an investment in the minerals sector as well as its subsectors, identify present and future bottlenecks in the industry, evaluate the impact of alternative taxation schemes and the regional consequences of an investment, and quantify employment and income changes associated with mining (Schena 1989; Rose and Miernyk 1986; Dorian 1987). Models based on inter-industry transactions can be used for planning, directing, and co-ordinating government investment programmes. Input–output analysis is a powerful way of analysing the probable effects of future minerals development on an economy when the results are used with proper caution.

In addition to input–output modelling, other techniques are available to assess the important role that minerals development has on an economy, though there is an insufficient number of such techniques and applications are limited. Moreover, China's blended economic system may not be appropriate for any particular modelling methodology. Like input–output analysis, other modelling techniques have specific advantages and weaknesses, with no individual methodology meeting all modelling purposes. Schena (1989) and Bosson and Varon (1977) explain that the void in techniques and their applications in developing countries in particular involves the institutional separation of geologists and mining engineers on the one side, and development economists and planners on the other. These specialists often possess different perceptions of development, and employ specialized terminology limiting communication between one another.

One modelling technique applied to a wide variety of policy issues including the impacts of a mining sector is the computable general equilibrium (CGE) model. In recent years, the use of CGE models has grown rapidly in the field of applied economic research. With a neo-classical foundation, CGE models are flexible enough to simulate imperfectly competitive behaviour, quantity or price adjustment lags, and widespread government intervention (such as that in China). The principal features distinguishing this generation of input–output-based models from the first generation Leontief-type input–output models are (1) the recognition that sectors in an economy are interdependent not only because they sell intermediate goods to one another (which is the sole source of interdependence in Leontief models), but also because all sectors ultimately compete for factors of production, and (2) sectors cannot supply unlimited amounts of output at fixed prices in response to demand stimulation (Cook 1984). Within the CGE model framework, as a sector attempts to expand output, it exerts pressure in factor markets, which usually involves changes in relative factor prices and, therefore, changes in relative commodity prices and incomes. Simultaneously, producers and consumers respond to the changes and create a new pattern of supply and demand.

The main advantage of CGE models is that they permit a simulation of a market system in which prices are the equilibrating forces, without necessarily dealing with a purely competitive economy. The model can simulate market conditions influenced by government policies and other exogenous forces. It is also suited to analyse policy choices involving exchange rates, tariffs, and so on, which operate through the price mechanism, and their effects on an economy. Being a static model, the CGE model does not represent a good forecasting tool.

One of the more widely used CGE models is the ORANI model of Australia, which successfully simulates a number of important features of the economy as well as some of the adjustment mechanisms that are associated with changes in the minerals sector. The ORANI model has been utilized extensively by Australian economists to quantify the effects of changes in the Australian minerals sector in such areas as investment, exports, and capital flows (Dixon *et al.* 1982; Cook and Lees 1984). Some of the findings using the ORANI model have contradicted findings using traditional Leontief input–output models,[10] resulting from a different set of model assumptions, equations, and characteristics.

An experimental CGE model for China was recently constructed at the East–West Center in Honolulu, one of the first ever published (Lee 1989). This model is a modification of the Yugoslav CGE model developed by Robinson and Tyson (1985). New features in the China model include explicit consideration of the existence in China of a dual allocation mechanism of central planning and markets associated with dual prices. The China CGE model takes into consideration the role of central planning as well as government intervention into markets. The model operates under static conditions but can be extended into a dynamic model for medium-term analysis. Some of the proposed applications of the China CGE model for policy analysis include (1) monitoring the expansion of the role of market forces on the Chinese economy, (2) assessing China's foreign trade and exchange rate policies, (3) evaluating the effects of reform on income distribution, and (4) analysing different sectoral allocations of labour. The model is also suitable for providing a disaggregated quantitative analysis of the impacts of additional investment by the minerals sector from additional mineral exports.

Another approach used in evaluating mineral resource development projects and programmes is benefit–cost analysis, which is a systematic method of identifying and measuring the economic benefits and costs of a project or programme. The benefits of a project are the values of incremental outputs of goods and services, including environmental services, made possible by the project, while the costs are the values of the incremental real resources used by the project (Hufschmidt *et al.* 1983). Benefit–cost analysis is concerned with the economy as a whole, and the welfare of a defined society. All project costs and benefits are necessarily discounted over time to make them commensurate. Benefit–cost analysis was developed in the United States during the 1930s to adequately assess the benefits and costs of proposed water-resource projects.

Today, while benefit–cost analysis is still considered a crucial part of all project appraisal, it is increasingly being accompanied by environmental impact evaluations. Most economists recognize the usefulness of benefit–cost analysis as a means of presenting the choice between competing uses of resources in a convenient and understandable manner. Though used widely in applied economic research, benefit–cost analysis has its limitations, including an inability to evaluate many non-market effects and intersectoral linkages.

In some scenarios, specific effects of a project do not impose a cost or yield a benefit within the confines of the project itself. Yet such externality effects may help a country or region achieve its economic objectives, and therefore should be included in any economic analysis. Unfortunately, however, externalities such as pollution, congestion, and health-related problems are often non-identifiable and, if identified, difficult to measure.

As emphasized in the input–output analysis presented in this chapter, many development projects as well as branches of industry are interrelated through forward and backward linkages to other projects or sectors of an economy. For example, a project may lead to higher prices for the inputs that it requires and lower prices for the outputs it produces. The project can also lead to lower demand (and prices) for competing projects or services or higher demand (and prices) for complementary ones. Such forward linkage effects thus may occur in industries that employ or process a project's output, and backward linkages in industries that supply its inputs. These linkages can lead to different demands of inputs and outputs and therefore create gains and losses for producers and consumers other than those involved in the project itself. Unlike input–output analysis, which is suited to capturing many of these multiplier effects, external costs (losses) and benefits (gains) may not be included in a project appraisal using benefit–cost analysis. Other external effects to a project to which it may be difficult to assign a cost or benefit are the international repercussions of a project development.

Examples of two recent applications of benefit–cost analysis in addressing questions relating to minerals development have included a study of the expected monetary benefits to Costa Rica from minerals exploration and development expenditures (Clark and Trocki 1991), and an evaluation of the costs and benefits associated with the construction of mineral processing facilities in the Philippines (Emerson and Warr 1982). In the Costa Rica study, the analysis served to quantify the economic impacts of mineral resource assessment programmes to determine the cost-effectiveness of resource evaluations and to ascertain future benefits of development to the nation. The benefit–cost analysis provided a framework from which to quantify the expected costs and benefits of mineral development activities to the nation. In the Philippines study, a comprehensive benefit–cost analysis was performed to assess economically the development of a large copper smelter. The study included an analysis of the important issues of foreign investment, taxation, and the choice of a discount rate. The study results had different implications for the advisability of building the smelter from any of the previous economic analyses.

Altogether, relatively few types of economic and mathematical modelling techniques are available to assess the impacts of mineral industry activities on an economy, and only a limited number of comprehensive applications have been completed. Those techniques available have specific advantages and disadvantages when compared to one another on an individual basis, but no single methodology is capable of fulfilling all modelling needs.

NOTES

1. Advantages of evaluating a centrally planned economy using input–output analysis include (1) the technique can be used to assure the internal consistency of an economic plan; (2) it can speed up the construction of such a plan; (3) after a balanced plan is calculated, the actual material flows from one sector to another could be easily and rapidly calculated through the multiplication of direct input coefficients and the levels of output of the corresponding products; (4) the spreading of technical progress could be aided by the use of planned direct input coefficients in the place of statistically derived ones; and (5) the feasibility of the planned set of input norms could be verified by checking whether the new [I-A] matrix is non-singular (Levine 1962).

   Along with these advantages, the input–output approach would entail some disadvantages including the loss of contact between planners and economic administrators during the balancing stage of plan construction (Levine 1962).
2. The method of valuation of output affects not only the recorded value of total gross output, but also the valuation of transactions in the cells of the transactions table. Two methods of valuation are generally recognized, namely the price paid by the purchaser, i.e. purchasers' prices or purchasers' values, and that received by the producer, that is, producers' prices or producers' values (Jensen and West 1986: 23–5). The 1981 input–output table of China analysed in this study is in producers' values. The difference between the two types of values is represented by the 'mark-up' or 'marketing costs', which include trade margins (both wholesale and retail), transportation (post-production) and storage, and commodity taxes (net of commodity subsidies). A table in producers' values shows marketing costs assigned in the tables as a direct cost to the sector consuming the commodity; if the table were prepared in purchasers' prices, these costs would be assigned in the sector producing the commodity (Jensen and West 1986: 25).
3. For a detailed critical review of the 1981 input–output tables of China, see Dorian (1988).
4. GRIMP (short for GRit IMpact Program) is a general utility input–output computer software package designed to update and implement existing input–output models. According to its authors, GRIMP was designed to promote regional input–output analysis from the level of simply a research technique to that of an operational planning

methodology. GRIT is an acronym for 'Generation of Regional Input–Output Tables', a procedure developed by R. C. Jensen and his research team at the University of Queensland for constructing hybrid or partial survey regional input–output tables. See Jensen *et al.* (1979).

5. Although the 24-sector China input–output table is capable of illustrating the interrelationships between the various sectors of the Chinese economy, the table would have to be adjusted for price differences to identify 'real' structural differences between China and other countries. That is, realizing that the prices of many commodities in China are substantially different from those in most other low-income countries, the input–output table would have to be corrected for these price differences if cross-country comparisons were to be made. Such price adjustments were made in the recently completed input–output analysis of China by the World Bank (1985*a*).
6. In general, input–output analysts prefer to assume that e.g. the household sector is an endogeneous component of the economy. Such an assumption suggests restrictive Leontief-type indifference curves, where prices have no effect on consumer final demand. An assumption that the household sector is an endogeneous component indicates that the level of local production is important in determining local levels of household income which, in turn, will be spent locally and consequently influence local consumption levels. An input–output model closed with respect to households contains a new closed matrix A, which is sometimes referred to as the augmented matrix. The inverse of the closed model includes an income multiplier.
7. For a discussion of the aggregate multiplier concept, see Dillard (1948: 85–100). Keynes developed the concept of an aggregate multiplier based on previous work by R. S. Kahn. He suggested that if a certain amount of income were injected into an economy, spending by consumers would rise by some smaller amount, and the extent of added income spent by consumers would become someone else's 'new' income.
8. State Statistical Bureau (1981–92 (1984; Chinese edn.): 270). In constant 1980 prices, labour productivity (yuan per worker) for state-owned industrial independent accounting units was 11,130 in 1978, 11,838 in 1979, 12,080 in 1979, and 11,863 in 1981. The 1981 decline in labour productivity may also have been due to a change in the composition of industrial output. Heavy industrial output actually fell in 1981, while light industrial output (which is labour-intensive) grew rapidly. This contributed to an overall slowing of industrial output growth, yet an increase in labour use.
9. Very little formal training in input–output theory and methodologies has occurred in China. None the less, economists are becoming increasingly aware of the usefulness of the input–output techniques, and efforts are now being made to employ input–output analyses to model price reform. Lu Nan and Li Mingzhe (1988) at the Price Research Institute, State Price Bureau, have been instrumental in modelling the impacts of price reform on China's overall economy. The Price Research Institute recently compiled a relatively large and detailed 1987 input–output table of China, although it was not released widely to the general research community.
10. e.g. when used to assess the impacts of additional investment by the minerals sector, the ORANI model concluded that new minerals sector investments would involve merely reallocating resources with very little change in aggregate real output or expenditure (Cook and Lees 1984). This contradicts the traditional Leontief-type notion that additional aggregate demand (such as additional minerals investment) produces significant multiplier effects, since the multipliers presume that other demand is not crowded

out and that demand stimulation will bring forward unused factor inputs so that aggregate output rises. Hence, when considering the impacts of additional minerals sector investment, the major differences between the ORANI CGE model and the Leontief input–output models are that some demand will be crowded out without additional capital inflow and all inputs cannot be automatically increased.

PART III

LONG-TERM PROSPECTS AND CONCLUSIONS

7
Future Mining Development in China

INTRODUCTION

In comparison to other economies in North-east Asia—Japan, the Republic of Korea, and Taiwan—China has a large minerals and energy resource base. The use of these resources in international trade following the adoption of the open-door policies in 1978 became a significant source of foreign exchange and also, in the petroleum sector, attracted a considerable degree of overseas investment. China's leaders, however, have discovered that China's resource supplies, in relation to its growing resource needs as an industrializing economy, are not inexhaustible. Per capita mineral resources in China are well below the world average. China is now in the process of adjusting from an economy once perceived to be resource-rich relative to demand to one in which many resources are in short supply.

Like that of its neighbouring economies, China's international economic future probably lies not in its export of raw materials but in its export of higher value-added manufactured goods. As inputs, mineral and energy resources remain fundamental to rapid economic growth, but the emphasis has changed. China must depend on external supplies to meet a growing share of its resource needs. Owing to relatively low efficiency in exploration, extraction, production, and transport, China's mining industry has not been able to keep pace with the rising consumption of raw materials and energy associated with strong economic growth. At the same time, given its resource base, China does have a significant opportunity to strengthen its resource position by improving domestic performance in production, delivery, and efficient use of mineral and energy resources. The further development of its resource base in an economically efficient manner is a critical challenge for the future. The deteriorating state of China's energy industry, for example, poses a serious threat to long-term economic growth unless output is maintained and efficiency improved.

PREDICTED/PLANNED FUTURE OUTPUT OF METALS, NON-METALS, AND MINERAL FUELS

Predicting long-term output levels of major mineral and energy commodities in China is difficult and risky. The Chinese government is capable of quickly changing the developmental direction and pace of its nation's mining industry with new laws or policy regulations, as was made evident in January 1989 when government officials banned the export of six selected metals. Although the bold decision was expected by some analysts, its potential impact on regional trade flows and long-term market conditions was essentially unassessable.

Since China's mining industry is linked closely to the nation's overall economy, it is important to examine prospects for future economic growth in order to predict possible scenarios for the country's mining industry. In 1991 China adopted both the Eighth Five-Year Plan (1991–5) and the Ten-Year Development Programme (1991–2000) with an objective of guiding the development of China's socialist economy, culture, science and technology, and people's living conditions to the end of the century. The plans envisage a much slower rate of overall growth than occurred during the Seventh Five-Year Plan period, and particularly a slower rate of growth in industrial output, at or around 6.5 per cent annually (Table 7.1). China's leaders will, however, find it difficult to slow the nation's economy, which has grown an average of almost 9 per cent a year in real

TABLE 7.1. *Targets under Eighth Five-Year Plan (1991–1995) and Ten-Year Development Programme (1991–2000), by industry branch (values at 1990 prices, and real % change, unless otherwise indicated)*

	1990	1995	% change[a]	2000	% change[b]
GNP (Rmb bn.)	1,740.0	2,325.0	6.0	3,110.0	6.0
Gross agricultural output (Rmb bn.)	766.0	878.0	3.5	1,043.0	3.5
Gross industrial output (Rmb bn.)	2,385.0	3,270.0	6.5	4,544.0	6.8
Tertiary industry	473.3	NA	9.0	NA	9.0
Grain (m. tonnes)	435.0	450.0	0.7	500.0	1.4
Cotton (m. tonnes)	4.5	4.7	1.2	5.3	1.6
Coal (bn. tonnes)	1.1	1.2	2.6	1.4	2.6
Crude oil (m. tonnes)	138.0	145.0	1.0	NA	NA
Electricity (m. kWh)	618.0	810.0	5.5	1,100.0	6.0
Steel (m. tonnes)	66.0	72.0	1.8	80.0	1.8
Rolled steel (m. tonnes)	51.5	59.5	3.0	NA	NA
Chemical fertilizer (m. tonnes)	90.0[c]	100.0	2.1	120.0	2.9
Railway cargo (bn. tonnes)	1.5	1.6	7.8	1.9	5.6
Yarn (m. bales)	24.5[d]	27.0	1.9	31.0	2.4
Chemical fibre (m. tonnes)	1.7	2.0	4.6	2.6	5.0

Notes: NA not available. Figures are rounded.

[a] Average annual % change, 1990–5.

[b] Average annual % change, 1991–2000.

[c] Effective components of chemical fertilizers in 1990 was 18.80 tonnes.

[d] The output of yarn in 1990 was 4.63m. tonnes.

Source: *Renmin Ribao* (1991).

GNP since reforms began in 1978. By 1994, China's economy is expected to be four times bigger than it was in 1978; if China continues to experience rapid growth, its economy by 2002 will be eight times larger than it was in 1978 (*The Economist* 1992). At that time, China will have equalled or exceeded the performance of Japan, Taiwan, and the Republic of Korea during their fastest quarter-centuries of economic growth.

While China's long-term economic plans clearly suggest a modest expansion of heavy industry during the 1990s, details on individual sectors are not readily provided. None the less, certain industry branches will be emphasized between 1991 and 2000, as indicated in Table 7.1. China is intent on vigorously developing its energy industry during the next several years primarily by boosting electricity output from 618m. kWh in 1990 to 1,110m. kWh in 2000. This would represent an expansion of electricity output of 5–6 per cent annually. Such an increase in power will result from expected gains in efficiency and development of selected major hydroelectricity projects and the two nuclear power stations under construction, at Daya Bay in Guangdong and at Quinshan in Zhejiang province.

As industry continues to expand in China, a readjustment of its structure is expected. Production capacity in agriculture will be increased to promote overall growth of the rural economy, while China's basic industries (energy, transport, post and telecommunications, and raw and semi-finished materials) and infrastructure will be favoured for increased public investment. China's processing industries and particularly those of metals and non-metals will be encouraged to adopt new technologies and methods, develop new products, and improve product quality to better meet the needs of the domestic and international markets. In capital construction, priority will be given to agriculture, water conservation, energy, telecommunications, and production of key products in short supply; in modernization, emphasis will be given to energy conservation, materials in short supply, improved quality and range, export products, import substitution, and industrial safety.

A few Chinese mineral-related organizations have published predicted or planned output targets for specific commodities to the year 2000. As was described in Chapter 5, the China National Nonferrous Metals Industry Corporation (CNNC), for example, has predicted that the total output of ten major non-ferrous metals would increase from 2.35m. tonnes in 1990 to 5.0m. tonnes by the turn of the century. Output of the metals reached 2.52m. tonnes in 1991. In addition to Chinese government forecasts, other projections have been made by Chinese and Western researchers, as well as Western government offices. Table 7.2 lists predicted or planned output levels for selected mineral and energy resources for 1995 and 2000. Many of these predictions are based on historical growth rates of production, recent capital investment levels in mineral-related industries, and planned government-funded mine construction. All predicted output levels in the table assume continued economic growth in China to 2000.

TABLE 7.2. *Predicted/planned future mining output of China*

Mineral	Predicted quantity (tonnes '000 unless otherwise noted)	Year	Principal source
MINERALS			
Chemical fertilizer	100,000[a]	1995	(1)
(standard units)	120,000	2000	(1)
Gold (tonnes)	155[b]	1995	(2)
Non-ferrous metals[c]	3,000	1995	(2)
(tonnes)	4,000	2000	(2)
Rolled steel	59,500	1995	(1)
Steel	72,000	1995	(1)
	80,000	2000	(1)
ENERGY			
Coal	1,230,000	1995	(1)
	1,400,000	2000	(1)
	1,365,000	2000	(3)
	1,600,000	2010	(3)
Crude oil			
Total	145,000	1995	(1)
	142,000 to 149,500	1995	(4)
	153,000 to 169,000	2000	(4)
Offshore	5,000 to 8,000	1995	(4)
	5,500 to 10,000	2000	(4)
Natural gas	20 bcm/y	1995	(2)
Natural primary energy[d]	1,172,000[e]	1995	(2)

Note: Figures are rounded.

[a] Effective components of chemical fertilizers in 1990 was 18.80m. tonnes.

[b] China aims to increase gold production during 1991–5 by around 40% compared with the preceding five-year period (Gold Field Mineral Services of London estimates that China produced approximately 396 tonnes of gold between 1986 and 1990).

[c] Includes aluminium, antimony, copper, lead, magnesium, mercury, nickel, tin, titanium, and zinc. To ensure continued growth of its non-ferrous metals industry over the next ten years, China's government recently freed a number of non-ferrous metal companies from mandatory state plans for production and sales, allowing them to make independent marketing decisions and enter international markets.

[d] Gross.

[e] Tonnes of standard coal. Represents an increase of 132 mt over 1990 production. Over the five-year period to 1995, the total energy saved nationwide will be equivalent to 100 mt of standard coal.

Sources: (1) *Beijing Review* (15–21 Apr. 1991); and United Nations Industrial Development Organization (1991); (2) *Beijing Review* (13–19 May 1991); (3) Johnson (1992); and (4) Fridley (1991*a*).

NEW DIRECTIONS FOR CHINA'S MINING INDUSTRY

China's long-term economic plans and prospects will be largely affected by developments in its mining industry. If the nation is to become a major supplier of a variety of mineral and energy resources, certain policy decisions and financial commitments must be made. Such policy choices range from the extent to which foreign assistance and co-operation will be allowed and encouraged to the desired level of minerals and energy self-sufficiency.

China's raw materials and semi-finished/finished mineral industries are heavy consumers of energy and new mines and processing facilities must therefore be rationally located in light of China's energy supplies and transport facilities. The nation's primary energy source is coal, which constitutes about 80 per cent of China's energy consumption. It is highly unlikely that this situation will change in the near future.

Currently, China's coal resources are clustered in five energy bases including Shanxi, Shaanxi, and Henan provinces, and Inner Mongolia and Ningxia Hui autonomous regions. By 2000, it is estimated that China's annual demand for energy will go up to the equivalent of 1.5bn. tonnes of standard coal, and the aforementioned energy bases will account for about half of the coal output. Any increase in coal production and utilization implies an increase in the volume of coal freight carried by all means of transport, which will be constrained by inadequate facilities. If China's industrial distribution remains as it is today, heavier loads will be placed on existing transport facilities at the beginning of the next century. In view of the transport problems, areas with supplies of energy resources must examine closely their plans for industrial growth.

Given the relatively large endowment of energy resources in the middle part of China, the State Council and other Chinese government bodies are aware that a shift in raw materials and metals production from eastern China westward may take place and perhaps should be encouraged to take place (Liu Jingtong 1986). Such a shift will probably improve China's geographic or regional distribution of industry. The recently announced 12-year exploration programme by China's Ministry of Geology and Mineral Resources emphasized that future prospecting activities will be shifted to the western part of the country since major energy and mineral resources in eastern and central China have already been mapped and are being mined.[1] China's Eighth Five-Year Plan promotes accelerating the pace of geological prospecting so as to ensure sufficient mineral reserves to facilitate continued economic growth.

Given China's immediate need for minerals and metals, however, the government has apparently decided to expand and renovate existing facilities in the east, rather than begin a gradual shift to the interior energy-rich areas of China. Many of the interior regions are extremely poor and plagued with limited transport facilities. Water shortages are another problem, so the Chinese government must take an active role in supporting the development of energy, transport, and raw material industries in these areas if a shift from East to West is to begin. Such support should go beyond financial assistance and supply of skilled personnel, and include government policies similar to those instituted along China's coastal provinces to invigorate their economies.

As a means of better co-ordinating future mining exploration and development activities in China, a law governing the exploitation of mineral and energy resources was formulated and adopted in 1986. On 19 March of that year the Mineral Resources Law of the People's Republic of China was adopted at the 15th session of the Standing Committee of the Sixth National People's Congress (Foreign Broadcast Information Service, 25 Mar. 1986). Although adopted in March 1986, this law was not effective until 1 October 1986. Three provisional regulations were adopted by China's State Council in April 1986 to facilitate the implementation of the law—'Provisional Methods on Registration and Administration of Minerals Exploration', 'Provisional Methods on Mining Registration and Administration by State-Owned Enterprises', and 'Provisional Methods on the Supervision and Management of Mineral Resources'. In accordance with the law, each Chinese province, autonomous region, and municipality has issued administrative regulations for mining by collective and individual mining enterprises in their own regions.

According to its text (see Table 7.3), the Mineral Resources Law was drawn up on the basis of the Constitution of the People's Republic of China for 'the purpose

TABLE 7.3. *Major provisions of the Mineral Resources Law of the People's Republic of China*

Characteristics

1. Adopted 19 March 1986; implemented 1 October 1986.
2. Consists of seven chapters,[a] 158 articles, and one appendix.

Major Provisions—Chapter I: General Provisions

1. Purpose of Law is to develop the mining industry, strengthen the work of prospecting for and exploring, exploiting, and protecting mineral resources, and ensure fulfilment of the immediate and long-range needs of the socialist modernization programme.
2. Mineral resources belong to the State. State-operated mining enterprises are the principal bodies to exploit mineral resources. The State encourages, guides, and helps the development of collective mining enterprises of villages and towns.

Major Provisions—Chapter II: Registration of Mineral Resources: Prospecting; Examination and Approval of Mining Applications

1. The State carries out a system of centralized registration of mineral resources prospecting. The scope of registration of mineral resources prospecting and measures for such registration shall be worked out by the State Council.
2. Applications for running State-operated mining enterprises should be separately examined and approved by the State Council, the State Council's departments concerned, and the provincial, autonomous regional, and municipal people's governments.

TABLE 7.3. (*cont.*)

3. Mining should be conducted in areas under State plans, in areas which are of importance to the national economy, and in areas where the State permits only protective mining of certain local special minerals.

Major Provisions—Chapter III: Prospecting for Mineral Resources

1. In conducting general surveys of mineral resources, the major mineral species should be surveyed, after which a preliminary and comprehensive evaluation should be made of geological conditions concerning the formation of paragenetic and associated minerals, as well as of the industrial prospects of mineral deposits in the area.
2. In prospecting for mineral deposits, a comprehensive evaluation should be made of paragenetic and associated minerals that have industrial value in the mining area, and their reserves should be assessed accordingly. Prospecting reports that fail to include a comprehensive evaluation will not be approved.

Major Provisions—Chapter IV: Mining of Mineral Resources

1. In exploiting mineral resources, it is imperative to follow a reasonable order of extraction, and appropriate methods of mining and ore dressing. Mining enterprises' rates of extraction and recovery, ore dilution, and dressed ore retrieval should meet the designed requirements.
2. In mining major minerals, unified plans and comprehensive efforts should be made to exploit and utilize paragenetic and associated minerals having industrial value in order to prevent waste.
3. In exploiting mineral resources, it is imperative to observe the laws on environmental protection, and prevent environmental pollution. In extracting mineral resources, it is necessary to make economic use of land. In case arable land, grassland, or forests are damaged as a result of mining operations, the mining enterprise should adopt locally suitable measures to reclaim the land for use, plant tress or grass, or take other measures to make the damaged land usable. If exploitation of mineral resources results in losses to other people in terms of production and livelihood, compensation should be made and remedial actions taken accordingly.

Major Provisions—Chapter V: Collective Mining Enterprises in Townships and Individual Mining

1. In dealing with collective township mining enterprises and individual miners, the State implements the principle of giving them active support, making rational planning, providing correct guidance, and strengthening their management.
2. Collective township mining enterprises and individual miners should raise their technical standards and increase the rate of recovery of mineral resources. It is forbidden to excavate ores indiscriminately and undermine mineral resources.

Major Provision—Chapter VI: Legal Liabilities

1. Those who carry out mining without proper authorization and mining permit, enter the mining area defined by the State or mining areas that are of great value to the national economy or other people's mining areas to carry out mining without authorization, or tap special mining resources that the State has designated protected mining should be instructed to suspend operation and pay for the losses. Mineral products and illegal profits should be confiscated and fines be imposed.

[a] Six of seven chapters covered in this table.

Source: Foreign Broadcast Information Service (25 Mar. 1986).

of developing the mining industry, strengthening the work of prospecting for and exploring, exploiting, and protecting mineral resources, and ensuring the fulfilment of the immediate and long-range needs of the socialist modernization programme'. The law is aimed at encouraging while regulating mining by state-owned, collective, and private enterprises. The law stipulates that no mining can take place without a state licence, which could be revoked if the law were violated, and it also requires mining ventures to pay a 'resource exploitation tax' and other fees to the state for extracting mineral resources—which are legally state property. The environment is protected in the new law by a special clause which requires mining enterprises to reclaim land that has been destroyed during mining.

Seven chapters and 50 articles are contained in the Mineral Resources Law, including chapters on general provisions, registration for mineral exploration, mineral exploration and exploitation, collective mining enterprises of villages and towns, and mining by individuals, legal liability, and supplementary provisions. According to its provisions, the state owns, controls, and supervises all mineral resources in China, and any firm desiring to explore for mineral deposits must register in accordance with the law with the State Council or a designated department. Development rights must also be obtained, which cannot be sold, leased, or pledged.

Article 4 stipulates that the state-operated mining enterprises of China shall be the principal force in the exploitation of the country's mineral resources. Collective mining enterprises of townships and cities are none the less encouraged to develop mineral resources within areas designated by the state, while individuals are permitted to exploit scattered and dispersed mineral deposits, as well as building materials.

Article 5 indicates that any firm which develops mineral resources in China will pay a resource tax in accordance with relevant state provisions. No further definition of a resource exploitation tax is contained in the law, however.

Before a mining enterprise is established in China, a government approval authority will review its application for design or mining plan, production technique, and planned safety and environmental protection measures. Approval is granted if an enterprise meets standards established by the Mineral Resources Law or relevant state provisions. The government will monitor all development activities to ensure that recovery rates in mining and processing are conducive to existing technologies. Article 28 states that by-product mineral ores with commercial value must be efficiently exploited and utilized as part of an overall development strategy, so as to avoid waste.

Chapter VI of the Mineral Resources Law describes the many legal liabilities and conditions related to mining in China. Mining rights can be terminated for many reasons, including operating without a licence, entering without authority into mining areas designated for the state, exploiting special varieties of minerals that the state had ordered to be put under protective supervision, and causing unnecessary damage to the natural environment. Violations of these conditions can be investigated for criminal responsibility in accordance with the provisions of Article 156 of China's Criminal Law. Smuggling or illegal selling of mineral resources is strictly prohibited and subject to severe criminal punishment.

In addition to the recently adopted Mineral Resources Law, other means of achieving current exploration and development objectives will be used by the Chinese government. These include better co-operation—both foreign and domestic. For China to implement some of its present mineral and energy policies successfully, foreign co-operation is crucial in areas such as technology transfer and generation of foreign exchange. The exploration activities called for in China's Eighth Five-Year Plan will require large amounts of money to be invested by both the Chinese government and foreign countries and companies. Likewise, activities initiated during the Ninth Five-Year Plan will probably require massive amounts of funds.

Besides foreign co-operation, lateral co-operation among China's mineral-related ministries is vital if China is to accomplish its objectives successfully. Recently measures have been taken at the ministerial level in China to expand lateral co-operation and exchange. The 1989 agreement among eight government bodies to intensify geological prospecting activities over a 12-year period is an example of improved co-ordination. This agreement was drawn up in part to end alleged interministry rivalries and, at the same time, eliminate duplication of exploration efforts. Today, SINOPEC and the Ministries of Supplies, Railways, and Communications are considering ways of jointly co-ordinating the nation's coal production, distribution, and pricing systems in order to curtail shortages and eliminate price gouging.

In oil, because of an increasing capital shortage nationwide, new petroleum refineries are no longer being developed by a single government body alone but, instead, by numerous energy agencies as well as local governments. Previously the central government had financed such operations. China's petroleum industry is now following a three-pronged strategy of pursuing joint investments domestically, equity investments in overseas projects, and the creation of joint Chinese companies involved in production, trading, or other activities. In 1992, for example, SINOPEC and SINOCHEM—the China National Chemicals Import and Export Corporation under MOFERT—jointly established the China Union Petrochemical Company to engage in crude oil export and import activities. Both SINOPEC and SINOCHEM hold 50 per cent equity shares of the new company, with the current president of SINOPEC serving as the Chairman of the Board of the new firm.

During the 1990s, the Chinese government can exercise several options in directing the nation's mining industry. To increase minerals and energy availability, China can: (1) expand domestic mines, oil fields, smelters, and refineries using capital investment; (2) search for and discover new mineral and energy reserves; (3) import more ores and concentrates or semi-finished and finished products;

or (4) invest in mines, oil fields, and/or processing facilities abroad. In view of China's Eighth Five-Year Plan, the government is at this time emphasizing options 1 and 2 as viable means to boost domestic minerals production, while also pursuing option 4 as a way to acquire foreign technology and managerial know-how. The country's massive requirements for imported quality steel products and manufactures, iron ore, chromite, alumina and aluminium metal, and selected fertilizers are considered extremely costly and aggravate China's balance payments.

As for option 4, China's overseas investments have continued to expand since the mid-1980s. Indeed, China has invested in more than 80 countries and regions worldwide since economic reforms commenced. The total number of China's equity or contractual joint ventures and wholly Chinese-operated companies in foreign nations has reached more than 560 (*China Daily*, 17 Apr. 1989). Though most of the overseas enterprises are profitable, about 10 per cent are running at a loss owing to lack of business experience abroad and use of outdated management methods.

In the minerals sector, Chinese leaders agree that for all minerals where deficiencies exist, the nation should deliberately seek to invest in mineral development projects within foreign countries through equity participation. The Chinese government implemented this strategy by investing in the Portland Aluminium Smelter in Australia (Victoria), one of the largest of its kind in the world. The 1986 investment by the China International Trust and Investment Corporation represented the largest Chinese investment in Australia at that time (*China Daily*, 30 June 1987). The smelter receives alumina from Alcoa's western Australian refineries and is currently undergoing expansion from 300,000 to 600,000 tonnes of alumina per year. The first and second potlines are completed and consideration is being given to the construction of a third. While China's investment in the entire project includes a 10 per cent equity participation, this may soon increase. Presently China receives 10 per cent of the alumina from the plant and is responsible for marketing this output.

The Chinese government invested in the Portland smelter for three primary reasons. First, the Chinese minerals industry is attempting to learn and acquire successful managerial skills, such as those used at many Australian mines and processing facilities. Secondly, China is in desperate need of alumina and aluminium, that is, the semi-finished and finished products. Thirdly, China is in need of advanced smelting technology and, by acquiring equity in a successful smelting operation, can obtain the knowledge needed to advance its own aluminium production industry.

Another large Chinese investment overseas is the Channar Iron Ore Mine, located in Pilbara in Western Australia. This mine was initiated by then Chinese Premier Zhao Ziyang and Australian Prime Minister Bob Hawke during Zhao's visit to Australia in 1983. The low overall grade of the Chinese iron ore deposits (only 5 per cent of reserves contain in excess of 35 per cent Fe) and the need to beneficiate them to obtain a concentrate averaging closer to 62 per cent Fe, combined with the availability of iron ore from Australia with a direct shipping grade of 63 per cent, prompted this joint agreement. Utilization of high-grade feed is critical if China is to maximize the efficiency and productivity of its technically advanced blast-furnaces. Moreover, imported iron ore used at coastal processing facilities would relieve much of the congestion on the country's rail system.

The first phase of the Channar Mine, in which China and Australia jointly invested about $250m., produced 3m. tonnes of iron ore in 1990, all of which was shipped and sold to China. Eventually, Channar shipments will reach 10m. tonnes a year. The Channar Mine will produce over 200m. tonnes of ore over 20 years, and thus is expected to be an important long-term and stable source of iron ore for China. The contract term is 30 years, including construction time (*China Daily*, 9 Nov. 1988).

Most recently, China is considering investing in another joint mining venture in Australia. China and Western Australia are reportedly undertaking a feasibility study in the construction of a gas-based direct-reduced iron plant, capable of manufacturing inputs for high quality steel production (Tse 1992). The proposed nearly $200m. processing facility would be located in Australia's Pilbara region, also the site of the jointly developed Channar Mine. Two steel plants in China are searching for supplies of between 200,000 and 500,000 t/y of direct-reduced iron.

Many Chinese organizations have been established recently with an objective of seeking joint-venture mineral opportunities abroad as well as investment possibilities. Such organizations include the China International Trust and Investment Corporation (CITIC), the China National Gold Corporation, and the China International Iron and Steel Investment Corporation (CIISIC). Senior officials in the China National Gold Corporation have visited several gold-producing nations in recent years in order to identify investment opportunities and promote the acquisition of processing technology. Mineral-related investments and joint-venture opportunities have been actively sought by the Chinese in the Philippines, Papua New Guinea, Chile, and Russia (copper), Brazil (iron ore), Japan (iron and steel industry technology and equipment), the United States (gold and petroleum), Canada (gold), Australia (iron and steel, base and precious metals), New Zealand (iron and steel), Turkey, Pakistan, and South Africa (chromium), and a host of other nations. In what was believed to be the first Chinese investment in Latin America, Shougang Corporation became the new owner of Peru's state-owned iron producer, Hierro Peru, in 1992 after offering $270m. for the long-term expansion of the facility.

Investment in overseas mining projects is not restricted to the Chinese minerals industry. In recent years China's petroleum industry has pursued joint ventures and other business deals covering upstream and downstream refining operations in foreign nations. In 1988 SINOCHEM signed an agreement with the Coastal Corporation to acquire for an undisclosed amount of cash 50 per cent interest in the corporation's United States West Coast refining/marketing business (*Oil and Gas Journal* 1988). Coastal operations include refineries, fuel oil terminals, and marketing outlets along the United States California coast. As called for in the agreement, SINOCHEM and Coastal recently negotiated a crude supply deal. Relations between SINOCHEM and Coastal began in 1978 when Coastal became the first oil company based in the United States to purchase and import crude from China.

With the passage of the 14th Communist Party Congress in October 1992, China's petroleum industry was further encouraged to expand ties with foreign nations. China, facing energy shortages at home, is looking to overseas mining projects for future supplies of oil and other resources. Companies are today clearly committed to internationalization, desiring to sell their expertise, technology, services, and products abroad. SINOPEC and SINOCHEM are reportedly considering forming a joint venture to pursue investment projects overseas (Fridley 1993). The China National Petroleum Corporation (CNPC), an upstream crude oil producer associated with the Ministry of Energy, recently set up operations in Canada to develop heavy oil deposits, while also investing in acreage in the United States along with US-based companies. CNPC is today considering joint exploration opportunities in upstream oil production in both Russia and Kazakhstan.

Option 1, which involves expansion of domestic mines, oil fields, smelters, and refineries is largely dependent on the availability of capital for investment. While China's relatively good credit rating was tarnished by events in 1989 and, as a consequence, downgraded by international credit agencies, foreign investment continues to flow into the country at record levels.

In the 1990s, China's credit standing is likely to improve as government officials appear fixed in their encouragement of foreign investment and efforts to build the confidence of overseas investors. Barring any domestic turmoil or violence, capital will probably become increasingly available to China. Chinese authorities have begun relaxing restrictions on domestic sales of locally made items that previously had to be exported. The huge domestic consumer market in China is clearly becoming more accessible. Investment opportunities are also being spread out geographically. In June 1992 Beijing authorized 21 additional cities—primarily along the Yangtze River Valley plus several in the north-east—to offer additional incentives to foreign investors (*Far Eastern Economic Review*, 24–31 Dec. 1992). The level of foreign loans to China for development projects will depend on the long-term stability of the Chinese leadership and prospects for China's economy. The country's future stability may hinge on the success of China's leaders in attracting foreign capital to interior, poorer parts of the country, as nearly 90 per cent of the cumulative foreign investments received by China have gone into the coastal provinces.

As for domestically provided capital for construction and expansion, owing to a general lack of hard currency and an overheated economy the Chinese government halted significant capital construction of most mineral projects during the recent Three-Year Economic Austerity Programme (1988–91), with the exception of the iron and steel and aluminium industries. In late 1988 the State Planning Commission had urged all localities and departments to take concrete measures to slow down the excessively rapid growth of industrial production. As part of this move banks were instructed to stop issuing loans to collectively owned establishments as well as village-run factories which compete for raw materials.

While some success was achieved in curtailing domestic capital outlays in 1989 and 1990, investments rebounded in many sectors in 1990 and particularly in 1991. Much of the steady growth of China's energy industry since 1979 has been attributed to massive investments in construction projects and technological transformation, though the returns on investment gradually declined.

China's investment in capital construction in energy industries increased yearly since 1981, as did its proportion of total investment. In 1991 the state invested 64.6bn. yuan into capital construction in energy industries, a remarkable 580 per cent increase from 1981 (9.5bn. yuan). Its proportion in the nation's total investment in industry rose from 21.4 per cent in 1981 to 30.5 per cent in 1991. Of this figure, 11.7bn. yuan were invested in coal-mining and dressing, a 405 per cent increase over 1981; 8.8bn. yuan were earmarked for the extraction of petroleum and natural gas, a rise of 280 per cent; and 3.8bn. yuan went to petroleum processing, an astounding 690 per cent increase from 1981. Through such levels of capital construction, China's coal-mining capacity increased by 241.6m. tonnes between 1981 and 1991. During the same period the nation's oil-extracting capacity rose by 140.7m. tonnes, and natural gas production capacity increased by 7.64bn. cubic metres. Though relatively small in volume, additional capacity is added each year through the technical updating and transformation of mining and processing equipment.

The second option—searching for and discovering new mineral and energy reserves—is already under way in China. New nationwide surveys of mineral and energy resources in China were begun in 1985 and continued into the Eighth Five-Year Plan period. As already mentioned, a long-term, comprehensive exploration programme was unveiled in November 1989, drafted by the Ministry of Geology and Mineral Resources and many of its departments (*China Daily*, 27 Nov. 1989). The programme was approved by the State Planning Commission and included in the State Council's general planning. To initiate the 12-year programme, the ministry had requested a 2 per cent share of the national budget—a 40 per cent increase over the previous 1.6 per cent share, but lower than its peak share of 2.06 per cent in 1957. It also appealed for more sources of investment, including financial aid from local governments and the pooling of funds for local prospecting work. Currently, the ministry is primarily dependent on appropriations from the central government, as two-thirds of its budget comes from the national treasury.

China is one of the very few countries of the world where there are still vast regions (particularly in the west) that have yet to be extensively explored and drilled. So it is probable that mineral and energy deposits will continue to be discovered at a rapid pace in the near future, though the size and quality of discovered deposits will decline steadily—barring the occasional exceptional and unexpected major find. China's greatest geologic potential for discovering oil deposits is in the western and northern onshore basins, particularly in Gansu, northern Qinghai, and central Xinjiang (Dorian and Clark 1987). Petroleum exploration (and development) will continue to focus on eastern regions for the remainder of this century, but in the long term both western and northern China hold promise of major oil discoveries. In view of the exploration results of the past two decades, it can be concluded that the best prospects can be expected in basins overlying old, stable platforms with thick unmetamorphosed carbon sections, such as the Bohai Bay basin complex, the Sichuan, Jungar, and Tarim basins (*PetroMin* 1989). Present onshore drilling is concentrated in the Shengli-Gudong, Liaohe, Dagang, Jungar, Tarim, and Qaidam basins.

Offshore China, too, has substantial oil potential, despite the disappointment of recent finds. To date, only 20 per cent of the more than 500 potential geologic structures identified by seismicity tests have been surveyed. In spite of intensive exploration since 1979/80 by about 50 foreign oil companies, China's current offshore output remains at less than 3 per cent of total production,[2] but is estimated to reach 110,000 to 200,000 b/d in 2000, or nearly 5 per cent of the total. Offshore production may peak in the late 1990s, followed by a gradual decline. The increase is based mainly on expected output from southern fields.

Option 3, which involves importing more semi-finished and finished products is a viable but costly option. Chapter 5 highlighted the extremely high costs to China of importing finished products of iron and steel, petroleum, aluminium, copper, and several other commodities. These imports are necessary because of a pronounced lack of advanced smelting and refining technology within China's metals industry, and a limited petroleum-refining system in the energy industry. Despite the massive strides in China's recent refining modernization programme, the nation's ability to produce certain petroleum-based products such as diesel and gasoline is constrained by limited internal transport of crude, intermediate and final products, and use of outdated refining technology.

With a general shortage of foreign exchange and an emphasis in both the mineral and energy industries on reducing commodity imports, exercising option 3 in the near future is unlikely. In all probability, the Chinese government will employ a combination of the various development options available to its mining industry. The options chosen will vary by industry and commodity, and depend on a number of economic and political conditions such as availability of hard currency, influx of foreign investment, inflation, and economic and industrial growth rates. In addition to the four options just described, there are indeed other strategies which China's leadership could implement in combination with these options. In China's energy industry, for example, conservation efforts are being boosted to curb wasted oil, gas, and coal. Energy conservation has included strengthening the administration of enterprises, renovating processing equipment, and improving the distribution of oil and coal products. Future efforts at energy conservation may involve replacing oil with coal whenever coal use is possible.

ENERGY ACCESS VITAL TO INDUSTRIAL DEVELOPMENT

The reform of China's economy, now under way for about one and a half decades, has resulted in significant changes in industrial strength, distribution, and concentration. Reform has, however, encountered many obstacles, including a pronounced shortage of capital, infrastructural constraints, and ineffective bureaucracy. Of critical importance to the continued revitalization of the Chinese economy and industry is the adequate provision of key raw materials, notably coal, electricity, and steel. As for energy, its access has always been held vital to industrial development, as China's economic structure is based primarily on the domestic supply of fuel resources. Energy exports have provided China with some of the foreign exchange needed to purchase overseas technology and equipment for construction projects initiated during the decade of reform. The Chinese leadership can no longer depend on huge energy export revenues as domestic consumption continues to increase while production stagnates and exports are declining.

As China's economic development proceeded during the 1980s, industries providing the raw and semi-finished materials required for industrialization grew in abundance. As this occurred, primary energy consumption rose by more than 40 per cent, with the greater part of this increase being met by additional coal consumption. Throughout the decade coal accounted for at least three-quarters of China's total energy consumption, and the situation cannot be changed in the short term.

Despite the recent successes in China's energy industrial development, the gap between energy supplies and demand remains a weakness in the national economy, hampering economic development. During the Seventh Five-Year Plan, gross industrial and agricultural output value increased by more than 100 per cent in five years, whereas primary energy (standard coal equivalent) rose by only 18 per cent and generated power increased by 39 per cent. Energy growth continues to lag far behind economic growth during the Eighth Five-Year Plan period. As domestic demand for energy rises in China, new efforts to promote efficiency in utilization and conservation are being implemented.

The shortage of energy, particularly electric power and petroleum products, is delaying the construction of some industrial plants and transportation systems. Power remains in short supply in China, particularly in rural areas. In many places, home electric appliances cannot be put to normal use owing to insufficient power supply, and occasional power cuts are not uncommon even in large cities. The failure to increase electric power capacity in concert with industrial expansion has led to persistent power failures which have plagued urban China. Because of the serious imbalance between energy supply and demand, it is estimated that more than 20 per cent of productive capacity is idle in China (Xin Dingguo 1988). In rural regions, biomass energy constitutes the primary fuel for daily needs, while commercial energy accounts for only one-fifth of the total. With an estimated 40 per cent of China's peasants living without electricity, a large-scale increase in the supply of commercial energy will be required if the government chooses to electrify rural areas (Xin Dingguo 1988).

China's energy problems have been a major concern for the nation's economic planners for years. None the less, industrial development will continue to be constrained by energy shortages, as no improvement in China's primary energy balance is expected by the turn of the century (Table 7.4). By 2000, the supply–demand imbalance could range between 80m. and 400m. tonnes of coal equivalent

TABLE 7.4. *Projected energy balance of China, 2000*

Demand/supply scenarios	Total primary energy[a] demand (btce)	Total primary energy[a] supply (btce)	Deficit (mtce)
SCENARIO ONE[b]			
	1.56	1.30[c]	260
	1.56	1.48[d]	80
SCENARIO TWO[e]			
	1.70	1.48[d]	220
	1.70	1.30[c]	400

Notes: btce billion tonnes of coal equivalent. mtce million tonnes of coal equivalent. Figures are rounded.

Median supply–demand gap is 220–60 mtce, or about 15% of total demand. This is nearly the same as the present shortage.

[a] Primary energy comprises commercially traded fuels only.

[b] Satisfies the requirement of quadrupling China's GVAIO between 1980 and 2000. Based on sectoral analysis, this scenario considers a change in economic structure, technical improvements, and energy conservation.

[c] Low case production of primary energy.

[d] High case production of primary energy.

[e] Postulates that the energy elasticity coefficient in China is 1, meaning that energy demand will quadruple with GVAIO by 2000 to 2.4 btce. Subtracting 300 mtce saved from increased technical efficiency and 400 mtce saved from structural changes, final demand is estimated to be 1.7 btce.

Sources: Fridley (1991*c*); Energy Program (1990).

(mtce) depending upon the effectiveness of energy conservation measures and the adoption of newer technologies. This likelihood has underscored the recent emphasis on energy conservation, improved productivity, and comprehensive utilization of resources, with all policies designed to realize the enormous potential of China's energy base.[3]

To enable the country to proceed with reform and industrialization, major emphasis was given to expanding the energy industry in the Seventh Five-Year Plan (particularly electric power), as well as increasing efficiencies in utilization. The Eighth Five-Year Plan similarly promotes active development of China's energy sector. The low efficiency of energy utilization in China is reflected in the high unit consumption of energy in proportion to China's gross industrial and agricultural output value. Plans have also been made to accelerate energy price reform, though government officials remain wary of the inflationary pressures likely to result from any price rise. At the recently concluded 14th Communist Party Congress, China's economic reformers reportedly announced that a tripling of the price of oil would occur in January 1993 to bring the rate more in line with international prices. China's poor efficiency in energy usage is partially the result of energy prices being set too low, which therefore provide little incentive to conserve fuels and promote high consumption levels.

China's electric power industry is receiving top priority in the government's current plans to rejuvenate the nation's energy industry. Since most new power plants in China are coal-fired, this fuel resource is also receiving high priority in the efforts to achieve a better balance between energy supply and demand. Concerns over coal are not about the identification of additional economic reserves, but rather the improved co-ordination, exploitation, treatment (washing), and transportation of the resource. Newly constructed thermal power stations in China are often unable to operate at near or full capacity owing to limited access to coal resources. Better co-ordination is needed between the electric power and coal industries to develop coal deposits that are linked to new power plants by adequate and reliable infrastructures.

The Chinese government has now adopted several policies that are aimed at ensuring the completion of its long-term energy production plan to facilitate the stable growth of the national economy and meet the demands for a higher standard of living. In addition to promoting increased development of thermal and hydroelectric power plants in China, government plans call for active development of the nation's coal industry, and expanded exploration for oil and gas resources. Though in recent years China has relied on increased investment in capital construction in energy industries to boost production capacity, the additional capacity generated per unit of investment fell throughout the 1980s.

To realize its long-term energy objectives, China must make the attainment of a balance in energy supply and demand the highest priority for capital construction, but seek productive investment opportunities in combination with measures to improve efficiency in energy utilization. As experience has shown, actively encouraging foreign investment is one effective method of raising funds for energy development. Other measures the government could consider in promoting its energy industry include, for coal, revamping transportation capacity and adopting new technologies; for oil, encouraging foreign investment and technology assistance; and for electricity, expanding the development of small power stations in rural areas. Ultimately, the adoption of a rational pricing system for energy resources will provide a much needed stimulant to energy industrial development.

In the next ten years and beyond, the reform of China's economic system will continue with an aim of promoting economic growth. Despite setbacks in the late 1980s, China is under intense international pressure to continue economic reform, though the effectiveness of such reform would be enhanced if China's political structure was also amended. To meet the demands of economic development, energy output must be expanded significantly to satisfy rising rates of consumption. It has been estimated, for example, that electricity production by the end of this century must be at least four times higher than in 1980, increasing from 300 terawatt-hours (Twh) in 1980 to 1,100–1,200 Twh in 2000 (Xin Dingguo 1988). This reflects the urgency of the energy problem in China and its potential inhibiting effects on industrial development. Expansion of energy output is not the only solution to China's energy problem, however, as improved demand management through price reform will also be required. Older, inefficient energy-consuming enterprises will need to be closed in an expanded effort to conserve the nation's energy resources.

FUTURE LABOUR PRODUCTIVITY AND MECHANIZATION

The future development of China's mining industry will depend not only on technological advances in old and big enterprises, but on improvements in medium-sized and small enterprises too. One means of assessing the likelihood of continued growth in China's mining industry is by examining labour productivity through time. In China an estimated 5 million people are employed at the more than 120,000 collectively and individually run mines, and millions more are directly or indirectly employed in the nation's mining industry. Thus mining is a source of large-scale employment and, with planned expansion in minerals and energy production, avenues for additional employment in China are likely to improve.

With an expanding mining industry, the demand for workers both skilled and unskilled is gradually increasing in China. In general there is a shortage of skilled workers and technicians in mining areas throughout the country. The mining industry is not as advanced technologically as in other industrialized nations of the world, as much of China's mining operations are carried out via manual labour and ageing equipment. However, recent trends in mining labour productivity in China are favourable.

Table 7.5 lists 1981–91 labour productivity figures for China's coal, petroleum, and metallurgical industries. During the Sixth and Seventh Five-Year Plans, substantial government funds were devoted to improving overall productivity in Chinese mining practices. At the same time large-scale renovations were undertaken in many of the medium-sized and large mining enterprises. Evidently these investments and efforts had little impact in the early 1980s, as suggested by declining productivity in the petroleum industry, and nearly steady productivity levels in the coal industry. By 1984, however, significant improvements in labour productivity were beginning to occur in both industries, as well as the metallurgical sector.

In 1991 the overall productivity of underground mining of metallic ore was 421 tonnes per person-year, an increase of almost 60 per cent over the 1982 figure. This is an important indicator of the change in productivity of the extractive segment within the metallurgical industry, as the majority of metal mines in China are underground. About half of China's metal output comes from such mines. Overall productivity of coal production also rose strongly during the 1980s, by 44.7 per cent. Labour productivity in crude oil production and petroleum-refining peaked in 1987 and 1986, respectively, before steadily declining through the early 1990s. Between 1981 and 1991, labour productivity of crude oil output fell by nearly 15 per cent, despite efforts by the government to improve the efficiency of production of the industry. Notwithstanding, with China's continuing efforts to improve the efficiency of mining operations, prospects seem good for additional gains in labour productivity in at least two of the three minerals-related industries.

Improvement of productivity in the years ahead will require increased mechanization, in addition to a safer working environment. Some of the many reasons why China's underground mining operations have failed to reach production capacity in the past include use of inaccurate geological data, over-ambitious initial designs, and low levels of mining technology and management expertise (Chen Yunhe 1986: 31–3). Because of China's ambitious plans for future minerals output, the design and development of new large and medium-sized underground mines will be in accordance with high mechanization standards. A 1986 study defined the various types of mines in China that should be seriously considered for mechanization—operating mines that have not reached capacity, large and medium-sized operations with relatively large and identified reserves, and mines that produce materials that are desperately needed for national economic construction or are potential export earners (Chen Yunhe 1986).

The future of non-ferrous metals output in China is largely dependent on improvements in labour productivity and mechanization. One of the major objectives of China's Seventh Five-Year Plan was to increase capacity of non-ferrous metals production, where more than 70 per cent of increased capacity was to come

TABLE 7.5. *Labour productivity in output terms of mineral and energy industries*

	1981	1982	1983	1984	1985	1986	1987	1988	1989	1990	1991	% change 1981–91
PETROLEUM INDUSTRY												
Overall productivity of crude oil production[a]	213	199.7	196.2	218	239.3	248.5	251.2	250	218.4	182	181.3	(14.9)
Refining[b]	NA	NA	824	864	928	938	923	931	881	885	831	0.85
COAL INDUSTRY												
Overall productivity of coal production[c]	0.870	0.873	0.891	0.903	0.939	1.001	1.053	1.092	1.157	1.233	1.259	44.7
METALLURGICAL INDUSTRY												
Overall productivity of opencast mining[a]	NA	3,445	3,483	3,824	3,902	4,321	4,943	5,028	4,878	4,758	4,957	43.9
Opencast pit miner[d]	NA	4,796	4,879	5,694	5,661	6,218	6,883	6,979	6,588	6,184	6,248	30.3
Overall productivity of underground mining[a]	NA	265	295	300	303	288	348	360	384	399	421	58.9
Underground miner[d]	NA	387	418	445	477	443	496	490	512	NA	NA	32.3[e]
Iron smelter, blast-furnace[d]	1,255	1,282	1,387	1,413	1,466	1,614	689	737	712	746	758	(38.1)
Steel making—open hearth[d]	515	506	534	527	490	564	577	591	599	600	587	14.0

Note: NA not available.

[a] Tonnes/worker.
[b] Tonnes/person year.
[c] Tonnes/worker shift.
[d] Tonnes/worker year.
[e] Inclusive of 1982–9 only.

Source: State Statistical Bureau (1981–92: Beijing).

from improvements in existing operations (Chen Yunhe 1986: 31–3). The remainder was to come from newly developed mines.

Recent figures indicate that more than 9bn. yuan was set aside for the technical renovation of the industry during the 1986–90 period, with two-thirds of all added production capacity being directly attributed to the technical reform (*Beijing Review*, 4–10 Mar. 1991). Improvements in technology clearly increased the strength of China's non-ferrous metals industry during the Seventh Five-Year period, as a number of key enterprises reached the world standards of the late 1970s and early 1980s with their updated technology and equipment. If China's non-ferrous metals industry is to continue to expand, overall efficiency of metals mining must also continue to increase. Two questions raised by this are (1) how much can labour productivity improve in Chinese mining operations, and (2) can an improvement in labour productivity result in a significant increase in metals output in the short term as well as long term? It is apparent from China's Eighth Five-Year Plan that the government is depending on increased efficiencies in mining, as well as processing and refining, to aid in boosting non-ferrous metals production in the country.

INTERNATIONAL PERCEPTIONS AND MISCONCEPTIONS ABOUT CHINA'S MINING INDUSTRY

In large part, foreign investment in China's mineral and energy industries is dependent on international perceptions of investment opportunities, benefits, and drawbacks in the country. The first two rounds of international bidding for China's offshore oil were relatively successful in attracting foreign companies, yet the third round stimulated little interest, leading to its premature closure. As Chapter 5 highlighted, there has been less than expected foreign investment in Chinese mining exploration and development activities, with the exception of Occidental's Antaibao coal project in Shanxi, and offshore and onshore oil and gas ventures.

China's non-fuel minerals industry has had a particularly difficult time stimulating foreign investor interest, despite strong efforts. This has occurred because of many factors. Initially, there is a general uncertainty about the nature of mineral deposits in China. Much of the literature in the Western press claims that mineral deposits in the nation are, in general, of lower grade and smaller than similar deposits elsewhere in the world. Some geologists dispute this perception as being based on definitional differences between mineral reserves in China (industrial and prospective) and reserves in the West (economic and inferred).

Another concern of foreign investors is the general lack of adequate and reliable information on tonnages and grades of prospective mineral deposits. That information which is available is often confusing because the methods by which the figures are computed are not defined. When describing mineral (or energy) reserves,

Chinese literature needs to define the ways in which such reserves were computed (that is, polynomial reserve computation versus kriging), and what in fact a reserve in China is. (Readers are referred to David (1988) for a detailed description of the various geostatistical ore reserve estimation procedures used today.)

There are several perceptions about China's mining industry that concern investors interested in either mineral or energy projects. One such perception is that minerals and energy development has been and will continue to be limited to the east (which in this case is defined as those areas of China east of a line running between Lanzhou, Gansu, and Kunming, Yunnan). It is estimated that as much as 90 per cent of the minerals and energy output of China comes from the more populated, more industrialized east. As a result, foreign investors tend to feel that it may not be worth while to invest in mining projects in western China because of the lack of infrastructure, and general lack of encouragement by the Chinese government to invest in the west. Indeed, up to and including 1992 Chinese government officials generally restricted foreign investment for land oil-field exploration and development to the ten southern provinces and regions—particularly northern Hainan Island, the Leizhou Peninsula, the Pearl River Delta, and certain other basins in the provinces south of the Yangtze River. Some active minerals and energy exploration has none the less taken place in Xinjiang, a frontier region rich in resources which probably will not be extensively developed until next century.

Two additional perceptions that foreign investors have about China's mining industry are that there are no laws or regulations guiding minerals and energy exploration and development in China, and that few laws exist regarding equity joint ventures using Chinese and foreign investment. These perceptions are slightly outdated, though it is true that the specificity of many laws to the mining industry is unclear.

From 1979, the year China began implementing its open-door policy, to 1993, the Chinese government has promulgated over 150 laws and regulations concerning external economic relations, including foreign trade, joint-venture enterprises, taxation, technology transfers, and special economic zones and open cities. Several other laws and regulations are now being formulated as well, which may directly or indirectly affect minerals and energy exploration and development, including corporation and bankruptcy laws. China's business laws can be broadly divided into five categories—laws related to joint ventures, laws and regulations related to taxation, laws and regulations guiding technology transfers, regulations concerning special economic zones and open cities, and laws related to external trade.

As a nation, China realizes that its mineral and energy resource development serves as a foundation for economic growth, as these resources are the necessary raw-material inputs for China's growing industrial sectors. Chinese government officials have thus also realized that further development of China's resource base will require foreign investment and technology, which is likely largely to involve joint ventures. Consequently, legislation does exist in China which governs the exploration and development of its minerals and energy base by Chinese and foreign investors. Indeed, a growing body of laws and policies pertaining to foreign investment and activities will be likely to affect the future of mining development in China. Some of the major laws and regulations pertaining to joint ventures and foreign investment in China are listed in Table 7.6.

The Regulations on the Encouragement of Foreign Investment, promulgated on 11 October 1986, reflect the more recent efforts on the part of Chinese government authorities to encourage foreign businesses to invest in China. With the investment environment in China being tarnished in 1989, several laws were passed in 1990 to improve conditions, including, firstly, an amendment to the 1979 joint-venture law, allowing non-PRC nationals to chair a joint-ventures board of directors and providing flexibility on joint-venture tenures; and, secondly, provisional land development regulations enabling foreigners to develop tracts of land in the open cities and special economic zones. The 1979 joint-venture law stated that projects must have a fixed term, and the rules provided a blanket 30-year limit (which was subsequently extended to 50 years). The 4 April 1990 amendments to the 1979 law lifted the 50-year cap on the term of most joint-

TABLE 7.6. *Major laws and regulations pertaining to joint ventures and foreign investment in China*

Law/Regulation	Date promulgated
Law of the People's Republic of China on Joint Ventures Using Chinese and Foreign Investment	8 July 1979
Provisional Regulations for Providing Loans to Joint Ventures Using Chinese and Foreign Investment by the Bank of China	13 Mar. 1981
Procedures for the Registration, Examination and Approval of Joint Ventures Using Chinese and Foreign Investment	24 Apr. 1981
Regulations of the People's Republic of China on the Exploitation of Offshore Petroleum Resources in Co-operation with Foreign Enterprises	30 Jan. 1982
Procedures for the Implementation of the Provisions for Labour Management in Joint Ventures Using Chinese and Foreign Investment	24 Dec. 1983
Accounting Regulations of the People's Republic of China for Joint Ventures Using Chinese and Foreign Investment	4 Mar. 1985
State Council Regulations on the Joint Ventures Balance between Foreign Exchange Revenue and Expenditure	15 Jan. 1986
Law of the People's Republic of China on Enterprises Operated Exclusively with Foreign Capital	12 Apr. 1986
Law of the People's Republic of China Governing Wholly Foreign-Owned Enterprises	12 Apr. 1986

TABLE 7.6. (*cont.*)

Law/Regulation	Date promulgated
Law of the People's Republic of China on Chinese–Foreign Co-operative Joint Ventures	13 Apr. 1986
Provisional Regulations on Chinese–Foreign Co-operatively Designed Construction Projects	5 June 1986
Provisions of the State Council of the People's Republic of China for the Encouragement of Foreign Investment	11 Oct. 1986
Implementation Measures of the Ministry of Finance for Putting into Effect the Preferential Terms on Taxation Provided in the 'Provisions of the State Council for the Encouragement of Foreign Investment'	30 Jan. 1987
Regulations on the Payment of Royalty for Exploitation of Offshore Petroleum Resources	5 Dec. 1988
Amendments to the 1979 Law of the People's Republic of China Regarding Sino-Foreign Equity Joint Ventures	4 Apr. 1991
Copyright Law of the People's Republic of China	1 June 1991
Income Tax Law of the People's Republic of China for Enterprises with Foreign Investment and Foreign Enterprises	1 July 1991
Patent Law of the People's Republic of China[b]	1 Jan. 1993

[a] By Jan. 1993 there were 20 major laws and regulations pertaining to joint ventures and foreign investment in China. In-depth discussion of each can be found in the sources listed below.

[b] First put into practice 1 Apr. 1985, the law was revised in Sept. 1992 by the Standing Committee of the National People's Congress. By the end of July 1992, the Chinese Patent Office had accepted 253,414 patent applications, 37,449 of which came from foreign applications.

Sources: *Beijing Review* (30 Nov.–6 Dec. 1992; 16–22 Sept. 1991; and 24–30 June 1991); Intertrade (1988); Jennifer C. Clark (1988); Ministry of Foreign Economic Relations and Trade (1987); Ross and Silk (1987); and China International Economic Consultants, Inc. (1986).

venture contracts, granting certain projects open-ended lifetimes. Though China's MOFERT officials have renewed their efforts to protect joint-venture interests by drafting new or revised laws in 1990, there remains considerable room for improving the uncertainty surrounding the legal environment.

As already discussed, the Mineral Resources Law of the People's Republic of China was passed in March 1986 and made effective on 1 October of the same year. Although the law does not directly address setting up mining ventures with foreign investment or through Chinese–foreign co-operations, it does lay down general principles that all mining activities in China must adhere to (Foreign Broadcast Information Service, 12 Mar. 1986). Questions dealing with foreign-operated or Chinese–foreign co-operative mining ventures in China will be addressed with the deepening of structural reform.

Indeed, in a move to help pave the way for increased foreign investment into the Chinese mining industry, in 1992 China's State Council charged the Ministry of Geology and Mineral Resources with the responsibility of drafting regulations to promote and manage overseas investment into mining (*China Daily*, 11 Jan. 1993). At a May 1993 four-day conference in Beijing, the long-awaited draft codes were revealed by government officials, defining the country's new approach to foreign investment into its mineral sector (*Mining Journal*, 4 June 1993). The regulations will guarantee the legal rights and interests of foreign investors, while linking the provisions of the general Mineral Resources Law of 1986, the Chinese–Foreign Contractual Joint Ventures Law, the Law on Chinese–Foreign Equity Joint Ventures, and the Law on Foreign Capital Enterprises. The codes, which consist of 35 articles covering a wide range of topics, are designed to supplement but not override provisions in existing laws. China's policymakers are reportedly considering offering financial incentives, such as tax cuts and simplified procedures, to encourage foreign investment into mining.

Although little has been written on the nature of energy and mineral joint ventures in China, the major legal framework surrounding the establishment of minerals-related joint ventures between Chinese and foreign parties has been documented by J. C. Clark (1988). In her paper, Clark stated that legislation pertaining to energy resources is embodied in the Regulations of the People's Republic of China on the Exploitation of Offshore Petroleum Resources in Co-operation with Foreign Enterprises[4] and the Rules Concerning the Levy and Exemption of Customs Duties and Consolidated Industrial and Commercial Tax on Imports and Exports for the Chinese–Foreign Co-operative Exploitation of Offshore Petroleum.[5] Analogous provisions for onshore energy and other non-fuel minerals are contained in the 1986 Mineral Resources Law of the People's Republic of China.

In addition to substantive energy and minerals legislation, environmental laws which will directly affect foreign investment in mining enterprises have also been promulgated in China. The nation's first environmental protection law was passed in September 1979.[6] This law is based on the 1978 Chinese Constitution, article 11,[7] which articulates that the state is responsible for protection of the environment and natural resources. The environmental law encompasses mineral resources and resultant by-products and consequences of mining ventures.[8] Specific provisions of the environmental law will prohibit indiscriminate mining practices and prohibit dumping of mining waste into rivers, lakes, or the sea. It remains to be seen what environmental protections and controls actually will be required of foreign mining enterprises.

Other regulations for the foreign investor to be aware of include the general foreign enterprise laws and regulations,[9] and Laws of Foreign Economic Contracts of the People's Republic of China.[10] Because several of China's joint-venture and foreign investment laws are new and untested, many foreign business enterprises and owners do not have a clear understanding of how they pertain to foreign economic co-operation and investment. Additionally, the imprecise applicability of the 1986 Mineral Resources Law to foreign joint ventures leads to some confusion.

In actuality, only a few joint-venture mining projects, such as the Occidental Antaibao mine, have gone to completion. And the distribution of sales income and other financial and economic conditions have tended to be negotiated on a case-by-case basis. Based on these and other attitudes, realities, and perceptions among the international community, foreign investors have generally believed that investing in China's mining industry would be risky. Uncertainty about the nature of mineral deposits in China, reliability of geologic and economic information, development objectives for western China, and laws concerning mineral and energy resource projects has dictated caution and conservatism to foreign investors. Because of this uncertainty, foreign companies tend to discount probable net revenues of a potential project, reducing the chances that it be undertaken.

PROBLEMS WHICH MAY LIMIT FUTURE MINING DEVELOPMENT

China has developed a multifaceted programme of minerals and energy development that is designed to substantially increase the production of many commodities for both internal consumption and export. Despite its efforts, however, China has struggled with a number of historical problem areas and constraints on development while adjusting to the open-door policy requirements to develop its mineral and energy resources. To a large extent, most of the major problem areas have been identified, and the central government continues to modify its development plans and legal and economic system in an attempt to resolve them. Seven areas of critical concern are divergent interests of state and local officials, peasant mining activities, transport bottlenecks, safety considerations and pollution, the regulated pricing system, excessive taxation, and the shortage of foreign investment in these industries.

1. The Local–State Dichotomy of Interests

Central government officials involved in mining activities often indicate that mines in China are developed primarily to serve the state, province, region, and county, and to make (some) profits as well. Although these motives are commonly cited, the priorities of many major mining projects seem to be to serve the county, region, and state—in that order—rather than the state, region, and county. On the one hand, state officials (for example within CNNC) generally view mines as suppliers of necessary raw materials to facilitate industrial development in China. In contrast, local government officials seem to view mines as sources of revenue and employment for local residents, including peasants. Mines in China can be thought of as social programmes, where employment of local miners and peasants is of utmost importance, perhaps even more so than earning profits. These differing interests/objectives, which were intensified by decentralization, lead to inefficiences in mine development and operations throughout China.

Conflicts between local governments and centrally managed ministries often arise over usage and control of physical infrastructure serving mines. It is common for local governments to build and operate rail links and highways servicing mining development areas in China. As a result, local government officials can charge ministerial organs, such as CNNC, fees for using the infrastructure. Such fees can be raised (or lowered) at the discretion of the local governments. Control over infrastructure sometimes means that local governments also have significant power over the distribution of revenues from a mining project. There tends therefore to be little incentive for increasing efficiency or adopting a uniform system of taxation.

In addition to rail links and highways, local governments also commonly provide electricity and water resources to mining projects in China, and charge central government organizations for their use. Local governments have been known to charge near-monopolistic rates for electricity and water services to mining areas, particularly once a mine or smelter has been put into operation. Because of this, central government organizations involved in mining operations are now trying either to purchase electricity or water resources, for example, or to establish long-term contracts with pre-set prices for such resources at the beginning stages of development of a mine.

2. *Inefficient Mining by Peasants (Farmers)*

Peasants mine in China for two primary reasons: (1) they are poor, and (2) mining is a quick way to earn substantial income. In addition, peasants believe that they have a right to claim mineral ores when, as is sometimes true, a peasant's family has lived in a mineral-rich area for generations. Although the potential source of wealth available to the peasants is a major incentive to mine, one fundamental reason why peasant or individual mining is today common in China is the decentralization mode that swept the country during the 1980s. Except for a brief period during the Three-Year Economic Austerity Programme, mining activities at the local and even individual level have been actively encouraged by the state as a means of gaining prosperity.

One of the major problems caused by peasant mining is the inefficiency of such practices. Most peasant miners have very little if any experience in professional mining activities, particularly underground practices. Additionally, peasants commonly mine the richest part of an ore body, while discarding as waste some of the lower-grade but still valuable ore (high-grading). Many Chinese government officials have repeatedly called for the use of regulatory measures to protect mineral resources from waste and damage (*China Daily*, 15 June 1987). An estimated

120,000 people have been mining in the rich Lanping lead and zinc mine in Yunnan province since 1985, inefficiently dividing many medium-sized and large mining areas into small ones for exploration by local enterprises and private miners (Foreign Broadcast Information Service, 25 Aug. 1989).

Despite being implemented more than seven years ago, China's Mineral Resources Law has not effectively combated the destruction and waste of state-owned resources. A fundamental problem is that peasants as well as collective-run enterprises scramble for valuable mineral resources, causing huge damage due to disorderly mining. A 1987 survey of state-run mines conducted by China's Ministry of Geology and Mineral Resources found that 70 per cent of them were hampered by illicit smaller pits in which many local officials held private shares. There were 400 state mining trouble spots at that time where illegal activities were under way, including the Gannan Wolfram Mine in Jiangxi province, Gejiu Tin Mine in Yunnan province, and the Xiaoqinling Gold Mine at the juncture of Henan and Shaanxi provinces (*China Daily*, 28 Feb. 1992).

In recent years China's recovery rate of mineral resources has ranged between 30 and 50 per cent, well below the world average and that of Brazil and India, which have mining operations on a scale similar to that in China. Huge mining losses occur in mining ventures controlled at the local and individual levels. In Shanxi province, which supplies a quarter of the coal consumed in China, recovery varies from 50 per cent in major state mines to 10 per cent in township pits. In 1985, when township mines produced 89.2m. tonnes of coal, a 10 per cent recovery meant a loss of 800m. tonnes of coal (*China Daily*, 27 Nov. 1989). Waste by peasant and self-employed miners also occurs when associated and usable minerals within an ore deposit are ignored or abandoned.

In addition to the problem of high-grading, peasant mining also leads to smuggling because of the lack of control and regulation by the state. Gold, for example, is heavily mined by individuals—as an estimated 300,000 licensed private gold-miners are registered in China, and countless more people are mining the metal without any permission. More than half of the nation's gold output is produced by small mines owned by local governments and collectives and by rural individual prospectors. And indeed, not all of this gold is sold to the state, primarily as a result of poor management. Gold-smuggling activities are rampant in China, due in large part to the wide gap between domestic and world prices. There were, for example, an estimated 4,200 gold-smuggling cases involving 8,800 people between the beginning of 1987 and the end of June 1988 (*China Daily*, 8 Nov. 1988). More than 640,000 grams of gold, 430,000 grams of silver, and 27m. yuan of illicit money were seized during this period. Tungsten and tin have also been smuggled out of China in relatively large quantities.

As a means of promoting orderly mining by farmers, the Chinese government tightened overall administration of mines operated by farmers during the recent economic austerity programme. In 1988 the Chinese government banned individuals from mining, smelting, and processing gold, and created the National Gold

Management Bureau to supervise all mining and processing activities. Licences held by individuals were temporarily withdrawn, and private mining activities were being encouraged at the collective level. At present, peasant mining activities are found mostly in pits abandoned by state-run mines or at the border of large mines.

Further efforts to regulate farmer mining activities have been made by local governments and relevant central government departments, which have been asked to allocate some money to help farmers operate and supervise mines. The Chinese government recently decided that developing mines will have higher priority in its use of poverty funds, where China allocates 1bn. yuan each year to countries suffering poverty. During the 1980s the Agricultural Bank of China loaned more than 4bn. yuan to farmers' mines even though the central government did not invest directly in them. In 1993 China's State Council began preparing detailed regulations to collect mining compensation fees from enterprises and individuals for use by the state in preventing high-grading and excessive consumption of resources (*China Daily*, 28 Aug. 1992). The fees were to go into effect sometime during 1993.

3. Transportation Bottlenecks

Owing to the great distances between many of China's mineral and energy bases and the industries that process or consume them, growth and expansion of these industries will require an efficient transportation system. Although China is well endowed in many major metallic, non-metallic, and mineral fuel resources, many are located in remote areas, adding to exploration, development, and transportation costs. Moreover, many processing facilities for non-fuel minerals and oil and gas refineries are located away from either extracting or consuming centres. In 1992, after state-imposed quotas were fulfilled, a number of state-run enterprises were forced to shut down either because of a shortage of raw materials due to transportation inefficiencies or to conserve energy.

As described previously in Chapter 4, nearly 85 per cent of proven crude oil reserves in China are located in the eastern part of the country north of the Yangtze River, and most of the probable (but not proven) reserves are in north and north-west China. The principal energy-consuming regions of China are in the east, north-east, and south-central parts of China, typically distant from energy-producing centres. As supplies of crude oil grew, refining capacity expanded as well, nearly doubling since the 1970s. However, owing to interministerial rivalries and local pressures, most new refineries were placed far from oil fields or easy transport links (Clark, Dorian, and Fridley 1989). Consequently, oil-rich regions have excess refining capacity and oil-poor but politically influential regions enjoy large modern refining complexes.

The distances between producing, processing, and consuming centres would not be nearly the impediment that it is to mineral and energy development if

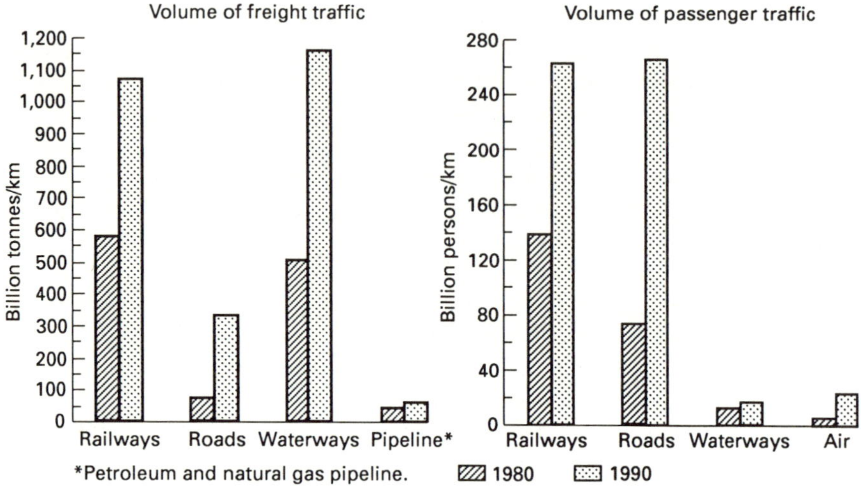

FIG 7.1. Recent transportation usage in China

China had a more modern and efficient transport system. In fact, the transportation of mineral and energy supplies places a heavy burden on an already overtaxed transport system. The backbone of China's transportation system is its rail network, which accounts for about half of total traffic (Figure 7.1). In 1991 the total length of railways in operation in China was 53,400 kilometres; less than 1,000 kilometres of new railway was added during the Seventh Five-Year Plan period. Coal and timber account for more than 40 per cent of rail freight, with iron ore, cement, chemical fertilizers, and semi-finished and finished metals accounting for significant and expanding shares. The lack of railway facilities has periodically halted increases in output at mines, particularly coal-mines. Many mines in China determine their output target not on the basis of their productive capacity but on the basis of the transport facilities at their disposal.

To improve China's coal transport sector, the state recently invested a total of 6.6bn. yuan, including 18.4bn. Japanese yen ($147m.) in loans, to build the Datong–Qinghuangdao railway in north China (*Beijing Review*, 4–10 Jan. 1993). This links the cities of Datong, a major coal producer in Shanxi, and Qinghuandao, the world's largest coal terminal. The 653-kilometre railway has the capacity to move about 55m. tonnes of coal annually. Its long-term handling capacity will approach 100m. tonnes per year.

Other efforts to relieve transportation problems have included the use of smaller coal-mines away from major mining centres, but these are typically not large enough to operate efficiently and have lower-quality deposits (Clark, Dorian, and Fridley 1989). Pipeline improvements have also shifted some of the oil transport burden away from the rail system.[11] More fundamentally, processing facilities should be sited according to economic rather than bureaucratic and political criteria. For example, as geological exploration efforts continue, non-fuel resource deposits are being discovered in western China. To avoid high transport costs, processing facilities should be located in this area and near energy resources such as hydropower, petroleum, or coal to facilitate energy-intensive smelting activities.

4. *Safety Hazards and Environmental Pollution*

The amount of ore mined by peasants in China is significant. Peasant mining is generally not supervised by state authorities; hence, there are no safety regulations imposed and few safety precautions taken. In general, peasants are nonchalant about their own safety as long as they can extract mineral ores. Accidents plague mining activities throughout China, contributing greatly to the inefficiencies of production.

In recent years the Chinese mining industry has gone through a formidable time in terms of safety. Almost 10,000 coal-miners were killed in accidents in 1991, a daily death toll of 27 (*Mining Journal*, 11 Dec. 1992). The shocking death rate—six deaths per million tonnes of coal—has prompted China to consider enacting its first mining safety law sometime in 1993. Compensation for injuries and deaths from accidents or occupational diseases is now costing the nation about 5bn. yuan ($1.35bn.) per year. Only traffic accidents in China have more fatalities than mining.

Unsafe mining practices are not necessarily restricted to non-supervised peasant miners. In 1988, for example, more than 4,000 petroleum/natural gas miners died in disasters—the worst year for such accidents since the founding of the People's Republic of China. Fatalities doubled from 1987, when 2,200 miners were reported in Chinese sources as killed. Workers in state-controlled facilities often consider safety of secondary importance, as fulfilling state quotas of output are of highest priority.

Random mining by peasants causes more than just safety problems. According to the *China Legal Journal*, rampant mining by peasants often threatens the health and safety of neighbouring residents because of randomly discharged waste water, gases, and residues (*China Daily*, 29 June 1987). Peasants invest virtually nothing in environmental protection. Normal production at state-run mines is often affected by indiscriminate mining, and surrounding areas are commonly polluted. Indiscriminate panning at the largest gold-mines in major rivers in Gansu province, for example, have caused serious soil erosion and pollution. The gold-mines, operated by more than an astounding 40,000 people, have also damaged vegetation, posing a direct threat to agriculture and human life.

Although peasants are often blamed for causing severe environmental damage in China, much of the nation's minerals industry, including state-run mines, is guilty of excessively damaging the environment. Environmental problems caused by mining can include subsidence, landslides, and mud flows, leading to heavy economic losses. For example, a survey in Yunnan, Guizhou, and Sichuan provinces in 1985 showed that the refining of sulphur was causing serious ecological damage, leaving areas of barren hillsides. The industry's total ouput value was only 50m. yuan, but the damage had been valued at 100m. yuan (*China Daily*, 26 Oct. 1985).

The total farming area occupied or destroyed by China's mining industry is estimated at 20,000 square kilometres (*Mining Journal*, 28 Aug. 1992). Economic losses resulting from natural disasters related to mining amount to hundreds of millions of yuan.

A major question regarding environmental pollution and China's mining industry is whether or not problems resulting from the pollution will seriously impede increased mining development in the future, particularly beyond 2000. Without a doubt, China has enormous environmental problems that threaten its ambitious modernization programme. A recent health study revealed that cancer and respiratory illnesses are among the leading causes of death in China. Since coal is the major fuel used in most places in China, it is the leading contributor to the growing sulphur dioxide levels in the atmosphere.

To coincide with the growing awareness of environmental problems worldwide, an effort is now under way by the Chinese government to develop adequate environmental protection strategies to tackle the deteriorating Chinese landscape. Recent efforts to protect the environment in and around mining areas are reportedly beginning to slow the growth of pollution in the nation. Officials from the Ministry of Geology and Mineral Resources have claimed that ecological disasters related to mining activities are dwindling, and previously disturbed farmland is now being restored. China's Eighth Five-Year Plan (1991–5) considers environmental protection as one of the country's basic national policies.

5. *An Irrational Pricing System*

Environmental pollution is, at least indirectly, related to the fifth major problem evident in China's mineral and energy industries—an irrational pricing system. One reason why pollution from mining and processing may be severe in China is that the costs of pollution are not internalized and reflected in resource values. To combat pollution, either prices need to reflect the additional costs brought on by environmental damage, or some other means of internalizing the costs—such as taxes or damage fees—need to be implemented.

China's commodity-pricing structure is considered one of the weakest aspects of its economy. Pricing is generally administered by the government at different levels. The price administration departments at province, autonomous region, and municipality levels can establish pricing principles, procedures, and standards within their assigned authorities. Prices have to be approved by the respective levels of government, and they can vary widely from one province or region to another.

There are currently three types of mineral and energy prices in China: (1) prices fixed by the state, (2) market-regulated prices, and (3) floating (or negotiated) prices. Prices fixed by the state are determined by the price administration departments or other authorities of government at and above county levels, based upon the price recommendation of enterprises or their management departments. Fixed prices are set with consideration for specific purposes, such as subsidizing consumption of necessities. They apply to mineral and energy resources sold within state quotas and may fluctuate within limited ranges in times of shortages or surpluses. Local prices of raw materials badly needed in production may be fixed at a level higher than nationwide state prices by provinces, autonomous regions, and municipalities, which may encounter difficulties if state prices are followed.

In China's oil industry prices fixed by the state are established at two levels—state-set high and state-set low—with the fixed higher prices approaching current market rates and fixed lower prices resembling international crude oil prices of 1982. With China's economic reformers now launching an ambitious effort to overhaul the country's struggling energy sector, an acceleration of price reform is likely whereby the state-set low price of crude oil will be dramatically increased or abolished. Officials from the China National Petrochemical Corporation (SINOPEC) recently disclosed the policy change, though it has not been acknowledged officially in Beijing. Government planners are evidently trying to determine how to prevent much higher oil prices from renewing inflationary pressures.

Above-quota mining output can be sold at prices which approach free-market levels, where prices are agreed upon by the buyer and seller with no restrictions. Market-regulated prices are generally fixed by individual enterprises in accordance with market conditions, to the extent permitted by laws and regulations of the state. Market-regulated prices do vary in different provinces.

The third type of price—a floating or negotiated price—is set between the buyer and seller of mineral or energy goods. Such prices are determined by enterprises according to the levels of fixed prices, differences of grades, profit margin, and ceiling and floor prices laid down by the price administrative departments or other government authorities. Floating prices generally range around the fixed price.

Up to August 1992, negotiated prices for above-quota crude oil in China were subject to a government-imposed price ceiling. By September, however, the price ceilings for 18 commodities, including oil, were eliminated altogether.

According to Chinese government sources, in the mid-1980s the percentage share of coal and petroleum products sold at fixed state prices, market-regulated prices, and floating prices were as follows (Ministry of Foreign Economic Relations and Trade Jan.–Feb. 1988): coal, 79.8%, 3.6%, and 16.5%; petroleum products, 42.6%, 0.5%, and 56.9%. (Prices of electricity and heat were sold in the following proportions—96.1%, 3.4%, and 0.2%.) By 1992 at least two-thirds of all petroleum goods sold in China's coastal provinces were marketed at above-quota prices. With price reform now being accelerated, most mineral and energy prices will probably be allowed to float freely in accordance with market conditions by around 1995.

China's economy is in disequilibrium today because of state-controlled prices and subsidies. In 1979 price reforms began in China in an effort to alleviate existing price imbalances, permit enterprises to earn profits at comparable rates, and allow prices to reflect supply and demand forces. With regard to resources, prices of mineral products, raw materials, and energy are, in general, too low and encourage waste. For the most part energy prices have remained fixed since the 1950s. To encourage industrial development, industrial fuel prices were set artificially low while households and other users paid two to three times the industrial price. By the 1980s, this pricing scheme had resulted in serious imbalances and widespread waste in energy use. Effective utilization of industrial energy had declined to around 25 per cent, nearly half the level in Japan.

Low prices for natural gas resulted in nearly one-third of the gas produced in some oil field being flared. Low crude oil prices (in comparison to final-product prices) encouraged the overbuilding of small refineries, many in provinces or localities far removed from sources of supply (Clark, Dorian, and Fridley 1989). From 1972 to 1988 subsidized oil was distributed at 100 yuan a tonne, equivalent to an average price of $6 a barrel at the exchange rate prevailing for most of that period (*Far Eastern Economic Review*, 12 Nov. 1992). By the beginning of 1992, it had been increased to 200 yuan a tonne, or slightly more than $5 a barrel at the present exchange rate. In effect, therefore, the price of China's crude oil has been declining in real terms for more than two decades.

Additional price reforms were implemented in 1983 when a multi-tiered pricing system for energy and mineral products was adopted. This allowed above-quota output from mines, refineries, and power plants to be sold at prices two to three times higher than the fixed state price. Although this measure stimulated production of some energy and mineral commodities, problems with relative prices remain. Above-quota prices for gasoline, for example, are still 30 per cent higher than diesel (on the international market, the gap is usually less than 10 per cent), encouraging investment in gasoline-producing equipment in refineries to maximize profit.

Unless completely addressed, the presence of distorted mineral and energy prices in China will continue to lead to imbalanced growth in output, and rational development of China's mining industry in the future will be difficult. Indeed China's mineral and energy industries may have difficulty becoming competitive globally if domestic Chinese producers are unable to respond to the same prices as their competitors. More fundamental energy price reform is now being discussed, but too little action has been taken to date. In view of the serious inflationary pressures arising from liberalization of the agricultural and industrial markets, China's leaders have been reluctant to rationalize and raise energy prices, which they fear will result in higher prices in virtually every sector of the economy. In a broader context, modernization of the national economy and economic growth in

China may be hindered if irrational prices remain much longer. At present, prices in China do not lead to an optimal allocation of resources based on scarcities; instead, government policies allocate resources with an intent to eliminate major differentials between geographic regions and classes of society.

6. *Excessive Taxation and Interest Rates*

The Chinese system of minerals taxation is unlike that of most other non-socialist countries, reflecting the unique character of China's economic, political, and social framework. In China, mining is perceived as part of the social welfare system and maximization of welfare payments, rather than profits, is encouraged under present revenue distribution schemes. Profits at most Chinese mining projects are therefore typically low.

Much of the revenues generated at an individual mine are returned to the state in the form of taxes. The Chinese income distribution system appears to be effective from a government perspective at capturing rents, and profit levels are carefully controlled. Taxes at typical non-fuel Chinese mines include (1) a products tax on sales revenues, (2) a state energy and communication tax, (3) a city construction tax, and (4) an income tax (which is only assigned after the loan payback is complete). During the 1980s, Chinese government officials were considering imposing a resources tax on domestic non-fuel mining operations, to be used to replace the income tax. With the adoption of the 1991 income tax, however, it is not clear whether such a tax will be introduced soon.

An important problem which may affect future minerals output in China is the unusually high taxation levels and interest rates (on loans) sometimes used in the minerals industry. A national strategy of income taxation for natural resource projects was first implemented in 1981 for selected enterprises. Non-fuel mining projects in China are, in general, heavily taxed, although taxes do vary from mine to mine and region to region depending on the profitability of the mine. Furthermore, these high taxes are in addition to operating costs which can also be high due to excessively high electricity, water, or transportation user fees set by local governments.

On the national level, two systems of revenue distribution are now in practice in China—a contractual responsibility system and an income-based taxation scheme based on a national income tax and a series of provincial and local taxes. The systems do not apply to joint ventures with foreign enterprises, although the distribution of income from domestic as well as Sino-foreign projects can, similarly, be decided by negotiation rather than legislation. Provisions in the 1991 income tax law of China are relevant to mineral joint ventures involving at least one foreign participant. As with domestically operated resource projects, special provisions are generally made for foreign investments in remote and economically poor areas, projects involving strategic minerals, or projects deemed necessary for national economic security. The provisions embody China's industrial policy in encouraging investment in key areas. They also follow the nation's ambition gradually to open up China's inland areas by extending links from coastal areas.

By 1984, all mining companies were to pay an income tax but the tremendous size and complexity of China's mining industry made the implementation of the system both difficult and confusing. The existing revenue distribution scheme provided little incentive for individual operations to improve efficiency. It was therefore temporarily suspended for many enterprises in 1987/1988, at which time a contractual responsibility system was introduced (Otto 1990). At this time several medium-sized and large mineral producers in China, including corporation and ministerial economic units, stopped paying the national income tax.

Under the contract-obligation system, each entity enters into a contract with the next higher level of management to produce or provide products or revenues equal to the amount produced or provided the previous year, with various economic incentives rewarding any increase over earlier levels (Otto 1990). Such a system of incentives was geared to boost the efficiency of the Chinese mining industry, which suffers from remarkably low efficiency levels in production and utilization. By mid-1988 nearly 8 per cent of China's industrial work-force comprised contract workers as a result of the fiscal tax and labour reforms instituted some months earlier (*China Daily*, 17 July 1988).

The revenue distribution system for major mineral projects in China not subject to the contract obligation system utilizes 13 primary distribution mechanisms (Otto 1990): (1) loan repayment; (2) products tax (pre-1989); (3) value-added tax (1989–present); (4) city construction surtax; (5) education surtax; (6) mine maintenance charge and subsidy; (7) depreciation allowance; (8) operating costs; (9) cost adjustment; (10) capital repair charge; (11) income tax; (12) excess profits tax; and (13) profit retention. This distribution scheme was applied to almost all large projects prior to 1987. Projects subject to income-based taxation have a priority to repay outstanding loans to the central government beginning in the first year of commercial production, if funds are available.

In addition to high taxation levels, interest rates on loans to mining developments in China are also high. In October 1986, the Chinese government increased the interest rate it charges on state-allocated loans to major Chinese mining corporations from 6.48 per cent to approximately 10.8 per cent for those loans being paid back within a ten-year period. For loan repayments taking more than ten years, the interest rate is even higher. (These interest rates are applicable to repayment of state loans but in practice the state is not usually capable of lending large sums of money to mining projects.) In 1988, with the introduction of the Three-Year Fiscal Austerity Programme, banks were further instructed to raise interest rates on financial loans to state and non-state industrial enterprises. The rise in rates was in addition to a reduction in the number of loans being offered at that time to enterprises.

As China continues to move towards a more market-based economy, efforts will expand to convert from a quota and price-controlled system to a more profit-driven and cost-minimization framework. Economic reforms will continue to influence the distribution of revenues from mineral projects. Since many such mining projects in China are not profitable, the central government will encounter serious difficulties in reducing or eliminating financial subsidies altogether. Many mining enterprises at the provincial and local levels exercise significant power, and will probably see little incentive to adopting a uniform taxation system. The untested nature and lack of uniformity of China's taxation system has tended to discourage foreign companies from participating in the nation's mining activities.

7. *Attracting Foreign Investment to Mining*

If China's mining industry is to advance significantly in the near future and beyond, investment from abroad is needed to fund mineral and energy projects. As in many countries, during the early 1980s most commitments of investment in the United States were for hotel and oil exploration, with the exploration turning up no significant discoveries. During the 1991–5 period, China is giving top priority to using foreign investment in, among other areas: (1) energy—this includes offshore and onshore oil exploitation, coal development, nuclear, thermal, and hydropower stations and related equipment, and power transmission equipment; and (2) the raw-materials industry—this includes non-ferrous and ferrous mines, the metallurgical industry, processing industries, steel rolling, and new-type building materials.

It appears that the Chinese government is resolved to attract foreign investment to fund mineral and energy projects. At this time China encourages various kinds of investment forms, but prefers equity joint ventures, especially export-oriented joint ventures. It is uncertain, however, whether China will be able to consistently attract the specific levels and kinds of foreign investment in minerals and energy that it requires to increase industrial output and efficiency significantly.

Investments in the oil and mineral sectors basically suffer from the same problems as other foreign investment in China—uncertainty about consistency in government policies and procedures, bureaucratic red tape, difficulties in obtaining foreign exchange to repatriate profits, rising costs of materials and wages, shortages and limited access to needed materials, and inadequate infrastructure. Foreign companies have been particularly disgruntled and disillusioned about investment in mining exploration and development activities because of the realization that high profits are not a sure thing despite a huge domestic consumer market and extensive geologic potential.

Part of this problem stems from a general lack of understanding about how the Chinese mining industry functions and, on a much broader scale, how China differs economically as well as culturally from other Asian nations and, even more so, Western nations. Economists from foreign nations have tended to evaluate Chinese mining projects according to Western financial and economic criteria. This is disturbing as China's mines are admittedly not the profit-driven operations that exist in the West. As already mentioned, mines in China serve as social programmes, employing hundreds or thousands of local residents. With an emphasis on labour-intensive operations, mineral and energy recovery and processing rates in China are, on average, much lower than in other parts of the world. However, the social benefits to government of employing large numbers of residents, many of whom are unskilled, have evidently outweighed the economic costs of lower efficiency. If foreign investors comprehend these and the many other different philosophies regarding the operations of a mine in China, frustration could possibly be avoided. Of course a foreign participant in any mining venture would have to be in a position to earn sufficient profits to make the investment feasible and worthwhile.

China's apparent willingness to adhere to its open-door policies is important to the country's overall credit rating. Though minor, recent amendments to the 1979 joint-venture law were considered a meaningful step in China's effort to improve the investment climate. In the short term, China's restrictions on foreign investment into mineral and energy prospects will continue to affect investor attitudes. However, recent efforts to open up domestic mines to overseas investors may pave the way for significantly greater foreign participation in joint-development projects. If further reforms are introduced in China with a recognition that direct foreign investment into mining, subject to reasonable policy regulations, leads to a beneficial arrangement for both the investing and host country, an increasing flow of capital to China is likely to be observed.

NOTES

1. A massive 12-year geological exploration programme was initiated in 1989 by several Chinese government bodies and approved by the State Planning Commission. The 1989–2000 study is a joint effort by the Ministry of Metallurgical Industry, China National Nonferrous Metals Industry Corp., Ministry of Geology and Mineral Resources, and five other state agencies. The emphasis will be a search for ores of aluminium, chromium, iron, lead and zinc, and manganese, and locating new deposits in the interior and western provinces of China. Exploration activities for industrial minerals required by China's chemicals and construction industries will include phosphate, potash, and soda ash.
2. Total offshore production in 1992 amounted to around 3.5m. tonnes, or 2.5% of China's total output of 140m. tonnes. About 2.5m. tonnes came from the South China Sea, primarily in several fields near the Pearl River mouth. An additional 1m. tonnes were produced by a number of small fields in the Bohai Gulf in northern China. The China National Offshore Oil Corporation expects three new offshore oilfields to come on stream in 1993, and a further eight oil/gas fields by 1996. CNOOC plans production of 12m. t/y of crude and 4bn. cm/y of gas from offshore fields by 1997 (*Petroleum Economist* 1993).

3. Although these policies were first promoted at the 11th Party Congress in 1980, they were heard with decreasing frequency over the ensuing decade. Since 1981, there has been an annual meeting of the Energy Conservation Working Group of the State Council, but no meeting was held in 1988. With Premier Li Peng's stress on energy conservation at the Seventh National People's Congress, the group was reconvened in June 1989, and the reports re-emphasized the need to give priority to energy conservation in the near term. China's Seventh and Eighth Five-Year Plans, as well as the nation's Ten-Year Programme, have reasserted the goals of pursuing energy conservation and improved efficiency as a means of reducing overall energy consumption.
4. Promulgated by the State Council on 30 Jan. 1981, and effective on the date of promulgation.
5. Approved on 28 Feb. 1982 by the State Council, and promulgated on 1 Apr. 1982, by the General Administration of Customs and the Ministry of Finance.
6. *Beijing Review* (9 Nov. 1979).
7. Constitution of the People's Republic of China, art. 11 (1978), as translated in Oceana Publications (1982).
8. The environmental law of China is fascinating in scope. However, its application is as yet uncertain. Like many of the substantive laws promulgated in China, the environmental law is largely aspirational in character. While any more than a general discussion is outside the scope of this chapter, the environmental laws are critical to any foreign investor wishing to develop mineral or energy resources in China because of the virtually inevitable environmental impacts. See generally Ottley and Valauskas (1983), *Beijing Review* (9 Nov. 1979), and Culliton (1979).
9. See Provisional Regulations of the People's Republic of China Concerning the Control of Resident Offices of Foreign Enterprises, promulgated by the State Council on 30 Oct. 1980, and effective on the date of promulgation; Notice of the General Administration for Industry and Commerce of the People's Republic of China Concerning Registration of Resident Offices of Foreign Enterprises, promulgated 8 Dec. 1980; Proceedings for the Registration and Administration of Resident Offices in China for Foreign Enterprises, approved by the State Council on 5 Mar. 1983, promulgated by the State Administration of Industry and Commerce on 15 Mar. 1983, and effective on date of promulgation; Notice of the State Administration for Commerce and Industry Concerning the Question of Registering Foreign Companies that Come to our Country to Co-operate in the Development and Contract for Projects, issued by the State Administration for Commerce and Industry on 12 Mar. 1983, and effective as of 1 Apr. 1983; Explanation of the Notice Concerning the Problem of Registering Foreign Companies that Come to our Country to Co-operate in Development and Contract for Projects, issued by the State Administration for Commerce and Industry on 4 May 1983.
10. Adopted at the Tenth Session of the Standing Committee of the Sixth National People's Congress on 21 Mar. 1985. A full translation of the text is contained in *Almanac of China's Foreign Economic Relations and Trade 1986* (MOFERT 1986–91: 516).
11. In 1991, the total length of petroleum and natural gas pipelines in China amounted to 16,200 km, compared to 6,700 km a decade earlier, or at the beginning of the Sixth Five-Year Plan period.

8
Summary and Conclusions

GAINING AN UNDERSTANDING

This book has examined thoroughly the growth of China's mineral and energy industries since the founding of the People's Republic, in terms of output, consumption, trade, employment, and overall impact on economic development. The study is intended to serve as a framework for understanding developments in China's mining industry in the context of central planning, and enable realistic projections to be made regarding the future outcome of the industry. It is hoped that the book can facilitate further analyses of the Chinese mining industry, which are highly recommended in view of the many fundamental policy questions confronted by the Chinese leadership, given the country's rapidly changing economic and political conditions, the growing value of the industry, and the increasing volume of statistical data now emerging from China. The role of mining in China's economic development is changing with rapidly rising GDP rates, technological progress, and marketization.

Since 1949, China's mining industry has evolved into an extremely complex entity in terms of structure, duties and responsibilities, foreign participation, and inter-organization co-operation and communication. Despite periodic setbacks and uncertainty, it has grown to become the world's fourth most valuable industry today. Though problems currently plague it, the industry is designed to ultimately increase output of many commodities for both internal consumption and export. As a consequence, the future of the world's mining industry will be significantly affected by developments within China, which simply can no longer be ignored in long-term global mineral or energy projections.

At present, the Chinese mining industry is largely insulated from international market conditions and regulated primarily by government policies and programmes. Increasing trade and foreign investment and technology assistance are beginning to erode the insulation, yet as long as an independent pricing system for mineral and energy resources exists in China there will be some level of protection. Untested laws and regulations regarding foreign investment and joint-venture activities have led to uncertainty and confusion among foreign investors considering Chinese mining projects. However, recent amendments to China's joint-venture and tax laws have provided foreign investors with increased access to mineral and energy projects, and preparations are now under way to further open the mining industry to overseas capital.

SUMMARY OF MAJOR FINDINGS

Mining has had a long history in China, dating back thousands of years. To gain a historical perspective on mining and the Chinese economy, long-term trends of cement, coal, rolled steel, and salt output were examined and described in Chapter 2. Such trends were influenced by a multitude of political, economic, and geologic factors, including industrial strategies, capital investment levels, and rates of discoveries of ore reserves. As industrial policies and political motives shifted from one five-year plan to another, the levels of government investment within the metallurgical, coal, and petroleum industries also tended to vary. Part I of the book dealt with the strong relationship between development planning in China and mining-related activities.

Chapter 2 revealed four distinct periods in the evolution of China's mining industry based on consistent output levels and relative stability within the extractive sectors. Period I, lasting from 1952 to 1960, was characterized by positive, rapid growth in output of major commodities until the disastrous effects of the Great Leap Forward gripped the industry. A proliferation of backyard steel furnaces eventually led to severe declines in efficiency and output. At around the same time the Soviet withdrawal from China further crippled the country's iron and steel industry, contributing to a significant drop in rolled steel output. Efforts at industrialization were futile at this juncture as production by a large number of units did not result in accelerated industrial development.

Unlike Period I, mining development during Period II (1960–71) was noted for instability and erratic behaviour. Shortly after the industry had apparently rebounded from the ill effects of the Great Leap Forward, the Cultural Revolution again slowed down mining activities. It was not until the first year of the Fourth Five-Year Plan (1971) that China's mining industry was mostly rejuvenated, led in part by expanded production of crude oil and refined petroleum products. Local programmes of rural industrialization, based largely on the iron and steel sectors, facilitated rapid industrial growth in China's countryside from the mid-1960s, which helped create the basis for Chinese industrialization until the early 1980s. Periods III and IV, which covered the years 1971 to 1980 and 1980 to the present, are both characterized by positive and strong growth in output of many mineral and energy commodities, and particularly rolled steel, cement, and refined petroleum products. Nationwide economic and industrial reforms during Period IV have changed the character of China's mining industry dramatically.

The system of central planning in China and its impact on historical minerals and energy development was the subject of Chapter 3. A complicated minerals and energy planning process was described in the chapter, where three planning tracks are employed in the development and implementation of policies. China's socialist economy is not guided by prices or other market forces, and therefore the system of planning must be evaluated to understand historical developments in the mining industry. From 1949, mineral and energy development strategies in China were typically based on domestic needs for industry, employment levels, technology availability, and export opportunities. Those sectors considered most crucial to China's overall economy, such as iron and steel, coal, and petroleum, generally received priority in China's five-year economic plans. Capital investment levels in these sectors remained high, even during periods of economic chaos such as the Cultural Revolution.

Today China is prominent in minerals and energy production, consumption, and trade, ranking among the world's leaders. The second part of this book examined the growing foreign economic relations that Chinese mining enterprises have established, and the overall impacts of mining on the Chinese economy. Specifically, Chapters 4 and 5 described thoroughly the minerals base and industry of China, and the significance of minerals and energy trade to the country. The various forms of foreign participation in China's mining industry were reviewed at the end of Chapter 5; these included joint-venture activities, direct foreign investment, and technology transfers.

Proven reserves of a number of mineral and energy resources in China are among the richest in the world; they include antimony, barite, coal, fluorspar, rare earth metals, tin, tungsten, and vanadium. Many mineral and energy deposits in China are located in remote regions, where difficult mining conditions and poor transport systems have led to relatively slow development. Chapter 4 highlighted the imbalance in China between mineral and energy supplies and their processing facilities and existing consuming centres. This imbalance has strained an already overloaded transportation system, and has hindered industrial development. As much as 20 per cent or more of China's productive capacity is idle as a result of insufficient power supply, as limited rail capacity sometimes makes it impossible to transport coal to industrial and urban centres.

Minerals and energy production and trade in China are handled by several ministries and corporations, employing more than 10 million workers throughout the country. With decentralization, the provincial and local branches of these organizations steadily increased their relative share of national output, exports, and imports. Non-state mining enterprises also thrived during the 1980s as a result of economic reform. In 1991, China was ranked number one globally in the output of raw coal, tungsten and cement, with 22.9, 5.2, and 18.1 per cent of world production respectively. Production continued to climb in 1992 for many commodities, particularly coal, cement, copper, gold, crude steel, nickel, petroleum, potash, salt, sulphur, tin, vanadium, and zinc. Since 1981, the beginning of the Sixth Five-Year Plan in China, output of raw coal increased more than 63 per cent, while production of cement rose by nearly 200 per cent. With the dramatic rise in urban and rural construction under way during the 1980s and early 1990s, China's cement production and consumption rose significantly.

The rapid expansion of China's economy during the 1980s and 1990s has boosted consumption of minerals, energy, and other raw materials. Tables 4.4–4.7

documented recent consumption levels of metals and energy in China, contrasting them with neighbouring Asian countries and the world. Between 1981 and 1991 China's rate of growth of consumption of nearly all major metals was much stronger than the world average, particularly for steel, zinc, magnesium, and tin. Similarly, rates of growth in energy consumption were stronger than world averages, though the growth rate of natural gas consumption in China was slightly below the world average. In terms of total primary energy forms, China's average annual growth rate of consumption was 5.3 per cent between 1981 and 1991, as compared to just 2 per cent for the world.

Owing to a lack of processing facilities and selected mineral reserves, China's recent rise in consumption has been met, in part, by foreign supplies. As China's foreign trade regulations were relaxed with economic reform, imports tended to grow much faster than exports. Chapter 5 examined trade flows of China in terms of value, destination or source, and future outlook. Nine commodity groups were evaluated, namely crude fertilizers and minerals, metalliferous ores and scrap metal, coal, coke and briquettes, petroleum and petroleum products, natural and manufactured gas, non-metallic mineral manufactures, iron and steel, non-ferrous metals, and manufactures of metals.

In 1991 China's mineral and energy trade resulted in an estimated $3.4bn. surplus, resulting largely from the high value of petroleum product exports that year—US$2.1bn. As for metals and non-metals, exports barely exceeded imports in China, as evidenced by a $500m. surplus in 1991. Iron and steel imports into China are extremely costly, accounting for a substantial share of the metals and non-metals trade deficit. As a result, if China is to improve its minerals and energy trade balance in the years ahead, the nation's reliance on imported iron and steel must continue to drop. This is particularly true in view of the growing domestic oil demand in China, which signifies that the country cannot depend on petroleum and petroleum product exports for foreign exchange very far into the future, barring a major oil discovery offshore or onshore.

To assess the economic interrelationships between the metallurgical, coal and coke-making, and petroleum industries and other producing and consuming industries in China's economy, Chapter 6 described an input–output analysis that was completed using the 1981 24-sector table of China published in 1987. In terms of output, the input–output model yielded total output linkage coefficients of 3.51 (coal and coke-making), 3.32 (metallurgical industry), and 2.27 (petroleum industry). The total output linkage coefficient for the coal industry, which is the largest coefficient among the three mineral-related industries assessed, indicates that each yuan of coal or coke sales to final demand, such as exports, generates an economy-wide increase in output of approximately 3.51 yuan. Of this 3.51 yuan, 1 yuan is accounted for by the direct increase in coal and/or coke production, while the other 2.51 yuan (flow-on effects) is the result of increased intermediate input purchases to produce the initial extra 1 yuan of coal output.

Chapter 6 revealed the significance of the mineral-related industries, and particularly the coal and coke-making industry, to China's overall economy in terms of economic interrelationships with and dependencies on other sectors. The specific nature of the interrelationships between the various mineral-related sectors will change as China's economic and industrial policies evolve. The coal industry purchases more inputs from more sectors, employs more workers, has larger output and income linkage coefficients, and has larger relative (personal) income rates than the other two mineral-related industries. Indeed, coal-miners in China are paid relatively high wages (considerably more than agricultural labourers, for instance), and they generally enjoy adequate housing facilities and benefits. Because of the oversupply of labourers in China, the government generally deems it socially beneficial to opt for production strategies which will make ample use of labour.

If it is assumed that from one year to the next coal and coke-making industrial exports increase by 134m. yuan while, at the same time, petroleum exports decrease by the same amount, the net effect on the Chinese economy is still positive. As discussed in Chapter 6, this example illustrates that the coal and coke-making industry evidently has a more widespread, positive impact on China's economy than does the petroleum industry. Because the petroleum industry of China buys its equipment and machinery from a relatively small number of manufacturers, the industry's interrelationships with other producing industries of China's economy are relatively few. In contrast, the coal and coke-making industry of China was determined to purchase more inputs from more sectors than either the petroleum or the metallurgical industries.

Although China's mineral and energy sectors have lower output and income linkage coefficients than agriculture-related industries, it is not necessarily true that sectors having larger coefficients are more socially beneficial. None the less, given the enormous magnitude of the agricultural sector in China, it is fully expected that its linkage coefficients would be larger than all other sectors in the country.

As in the past, the future of China's mining industry will depend largely on the economic plans of the nation and major political developments. A variety of problems and constraints, described thoroughly in Chapter 7, will also help shape the future of the industry. China's national economy will continue to be characterized as one with chronic shortages of energy, which will impede many new mining development activities. Other problems include inadequate transport systems, irrational pricing policies, and inefficient peasant mining practices. Environmental pollution from mining is another problem, and now considered a threat to the nation's overall well-being. Overexpansion of the mining industry in certain parts of China has led to serious pollution problems, including deforestation. Large amounts of resources and energy are consumed in China's mining, processing, refining, and manufacturing sectors. At the same time, huge quantities of solid wastes, waste water, and waste gases are released into the environment.

Chapter 7 also examined the numerous perceptions and misconceptions among the international community regarding China's mining industry. Importantly, these perceptions have tended to contribute to general scepticism about the industry, and to relatively little foreign investment into mining. Uncertainty and lack of knowledge about China's mineral resource and reserve definitions, reserve estimation procedures, mineral resource laws, and joint-venture regulations have, without doubt, dampened some overseas interest in investing in Chinese mineral or energy projects. This is unfortunate, as it was the hope of the Chinese government in 1979, when the nation took a historic step of welcoming foreign joint ventures, that direct foreign investment in China would aid development.

China's current Eighth Five-Year Plan (1991–5) and Ten-Year Development Programme (1991–2000) address some of the problems defined in Chapter 7, though capital constraints may limit the government's ability to build or improve infrastructure and combat environmental pollution. China recorded its largest fiscal deficit ever in 1991. The nation is, however, committed to an acceleration of geological prospecting activities, with an aim of discovering new deposits of minerals in short supply. The long-term plans emphasize technical renovation of metallurgical and petrochemical operations, in order to boost overall output and efficiency. Recent plans to open domestic mines to foreign investment may, when implemented, dramatically boost overseas interest in China's minerals and energy industry given its size, variety, and potential.

PROSPECTS AND CHALLENGES AHEAD

Nearly a decade and a half has passed since China initiated wide-scale industrial and agricultural reform, including the open-door policy towards foreign investment and technology acquisition. Reform is an ongoing process and, although considerable progress has been made, difficulties and setbacks have arisen.

Several developments during the 1980s and early 1990s either brought about or will bring about changes in China's longer-term minerals and energy policies. Decentralization dramatically changed the levels of authority of state, provincial, and local mining enterprises. With decentralization, provincial/local enterprises become wealthier and more autonomous, and better capable of dealing directly with foreign companies. But decentralized authority also led to problems of corruption, price-gouging, and excessive competition among provinces. For these and other reasons, the Three-Year Austerity Programme initiated in 1988 intended to recentralize some control over production and trade enterprises while restoring order to the Chinese economy. Despite these efforts, the Chinese economy continues to expand at a rapid pace today, and mining enterprises are gaining more autonomy in decision-making and dealings with foreign enterprises.

The past several years have helped Chinese leaders realize that despite a huge resource base, growing domestic demand will absorb an increasing share of production and necessitate increasing imports. This is placing enormous strains on an already overstretched metals industry. Although generally improving, China's iron and steel and non-ferrous metals industries continue to be plagued by deficits. And with growing domestic oil demand, the country cannot continue to rely freely on petroleum and petroleum product exports to fund industrialization and provide one-quarter of China's foreign exchange earnings. This was made clear in 1986, and is sound policy in the light of volatile global oil prices and uncertain future onshore and offshore output levels.

Another concern raised in recent years that is likely to affect future mining strategies in China is the general lack of foreign investor interest in major mineral projects and, more recently, in offshore oil bidding. Clearly, China's drive to attract foreign investors has encountered difficulties, and the civil unrest of 1989 temporarily exacerbated this situation. The principal worries of overseas investors arise from three uncertainties—the stability of the Chinese leadership, the consistency of economic and industrial policies, and the outlook of China's economy. Yet, to further develop its mineral and energy resources (particularly offshore oil deposits), China will require large quantities of Western exploration and drilling technology. The country will not, however, become dependent on foreign technology, but instead will attempt to rely ultimately on its own equipment.

Foreign investor concerns particularly relevant to the Chinese mining industry include untested laws and regulations, unscheduled rises in prices of raw materials, corruption, and inadequate data on reserve/resource inventories. Though difficult, foreign companies need to weigh the potential problems of investing in China against the long-term outlook of an expanding economy containing the largest consumer market in the world.

Since 1978, real per capita gross national product (GNP) has more than tripled in China. In 1992, GNP rose by more than 12 per cent, foreign trade by over 22 per cent, and domestic investment in fixed assets by greater than 40 per cent. If China achieves its economic goals for the year 2000, the nation's economy will be eight times bigger than it was in 1978 (*The Economist* 1992). China's estimated aggregate GNP in terms of purchasing power is already greater than that of the United Kingdom or FRG. If the country's economic performance in the next two decades equals that since open-door policies were first implemented, China will become the largest economic power on earth. While these developments are still, of course, speculative, they emphasize the magnitude of the changes which lie ahead, and the need to begin to adjust investment strategies and goals now.

China's economy is today moving through a transitional phase, from being a socialist economy of greater than a billion people which, in the past, followed isolationist tendencies, to a socialist economy with newly instituted market forces that is opening up to the world and relying on foreign investment and technology. In such a transitional phase, political disruptions are expected periodically as well as political achievements; the transition is, by its nature, subject to volatility. Moreover, China's transition does not involve only the nation's economy, but also its political, social, and cultural systems. Imbalances and inconsistencies between these systems are likely to continue to varying degrees contributing to volatility and, in turn, uncertainty. China today is at an ideological crossroads, where traditional, Confucian ideas are rivalling Marxist-Leninist theories and newly introduced Western philosophies. The student/military confrontation in June 1989 was an outgrowth of the tension and turmoil associated with a transitional phase. The student demonstrations of the spring of 1989 were in actuality part of an evolution of vocal unrest that began in 1986.

Given that China is moving through a transitional phase of a length that is not predictable, the likely characteristics of the Chinese mining industry in the long term are not easy to define with a high degree of precision. However, patterns of historical development and existing conditions and constraints allow plausible scenarios to be constructed. Clearly, the nation is well endowed in mineral and energy resources, though China is struggling with a number of historical problem areas and constraints in development while adjusting to the open-door policy requirements to develop its mineral and energy resources. China's mining industry already ranks high according to world standards but still operates at well below its potential, and the country has a very low per capita consumption of metals and minerals. The 1990s should be a time for expanding market incentives and seeking foreign assistance if government officials are anxious to further develop the mining industry in a rational and co-ordinated manner.

Despite encountering numerous problems, mining has advanced considerably in China over the past four and a half decades. Indeed, such advancements have come about with relatively little foreign investment, particularly during the isolation periods of the 1960s and 1970s. This raises the question as to what level of foreign investment is really needed to facilitate growth of the mineral-related industries in China. While the magnitude of foreign investment into mining is important, the nature of the targeted projects and activities is also critical.

If the Chinese mining industry expands its level of co-operation with and competition against foreign companies, the industry will become more closely in tune with international markets and thus affected to a greater degree by changing market conditions. The adoption of international market prices would also tend to increase the extent to which the domestic mining industry of China becomes responsive to worldwide market conditions. This implies that in the long term setbacks in China's mining industry may not be as severe as they have been in the past, as domestic factors will not be the only ones influencing the industry. Unlike market forces, central government plans can affect an entire industry immediately upon implementation.

Today, certain aspects of the Chinese mining industry are very promising. Mineral and energy resources are plentiful in China and have not been fully exploited or utilized. Resources of nearly 160 different kinds of mineral and energy resources have now been identified, while the proven deposit sizes of many resources in China rank very high in the world. The Mineral Resources Law of the People's Republic of China adopted in October 1986 is helping to unify planning and management of prospecting for mineral resources, and industries and departments which have unnecessarily duplicated each other's work. Future laws concerning minerals exploitation in China will probably contain specific references to foreign company involvement, perhaps reducing their uncertainty and confusion about important legal issues.

Although foreign investor interest in China's offshore oil prospects floundered during the third round of bidding, the recently adopted Regulations on the Payment of Royalty for Exploitation of Offshore Petroleum Resources offer companies more relaxed guidelines and an easing of restrictions. Recent Chinese incentives to foreign oil companies in hopes of stimulating their interest during the latest bidding round have included a lowering of royalty charges, an easing of drilling obligations, and reduced requirements to train Chinese staff. In 1992 China opened the East China Sea to foreign investors for the first time ever. Efforts to enhance foreign interest in onshore oil exploration and drilling activities are also likely to increase, with China's largest geologic potential for undiscovered petroleum being in the western and northern basins.

As a result of increased research and development, technology acquisition, and capital investment, labour productivity in the coal, petroleum, and metallurgical industries rose significantly after 1985, offering promise that further technological advances made in the 1990s will be equally far-reaching. Indeed, through technological transformation during the Seventh Five-Year Plan period, China's steel-smelting capacity rose by 11.2m. tonnes, extraction of petroleum by 4.1m. tonnes, and output of coal under state control (distribution) by 16.6m. tonnes. With increased research and development and technology imports likely in the near future, the production and processing capacity within China's mining industry may continue to expand at some level through renovations and advances.

The ultimate direction that the Chinese mining industry follows will depend, at least in part, on the pace of decentralization and reform within the industry. As emphasized throughout this book, central planning and political developments in China since 1949 have had a profound influence on minerals and energy development. In general, when economic and, more specifically, mineral or energy development plans were formulated in an orderly and realistic manner, such plans were by and large implemented successfully. To the contrary, when impractical industrial plans were made, output targets, for example, may have been met initially, but eventually problems plagued production. This occurred between 1958 and 1962 when the Great Leap Forward and People's Commune Movement called for accelerated economic and industrial growth. Although industrial output levels, including minerals and energy, increased dramatically until 1961, by 1962 wear and tear on machinery, use of outdated technology, and inefficient mobilization of China's labour-force led to havoc in the nation's economy and industrial sector.

Economic reforms in the 1990s have further decentralized China's mineral and energy industries, as direct central planning of output, distribution, and even prices has been significantly reduced. While domestic investment into mining is still controlled largely by the State Planning Commission, efforts are increasing to allow local enterprises to utilize foreign capital. Since December 1978 the centrally planned economy of China has gradually yielded to a more blended economy where political and economic decisions are guided by both central planning and market forces. Major developments of socialist reforms have included permitting a diversified economic system, introducing private ownership while ensuring that state ownership remains dominant, and allowing market forces to govern prices while retaining a reduced level of central planning. If indeed the nature of central planning in China continues to change in the years ahead, this, in turn, will have an impact on minerals and energy development in the country. The extent of the impact will only be revealed with passing years.

BIBLIOGRAPHY

Australian Journal of Mining (1991), 'China Digs up a Big'un', 6/62 (Nov.): 57.

Beijing Review (4–10 Jan. 1993), 'Energy Shortage Hinders Economy', 36/1: 5–6.

—— (30 Nov.–6 Dec. 1992), 'New Patent Law Promotes Opening', 35/48: 18–20.

—— (10–16 Aug. 1992), 'China Strives to Introduce More Foreign Capital', 35/32: 22–6.

—— (16–22 Sept. 1991), 'Copyright Comes under Legal Protection', 34/37: 23–5.

—— (24–30 June 1991), 'Steel Industry Reaching World Standards', 34/25: 20–3.

—— (13–19 May 1991), 'Excerpts of the Eighth Five-Year Plan (1991–1995)', 34/19: 11–16.

—— (6–12 May 1991), 'Development Blueprint for 1991–2000', 34/18: 18–19.

—— (15–21 Apr. 1991), 'Report on the Outline of the Ten-Year Programme and of the Eighth Five-Year Plan for National Economic and Social Development', 34/15: pp. i–xxiv.

—— (4–10 Mar. 1991), 'Nonferrous Metals Going International', 34/9: 17–21.

—— (7–13 Jan. 1991), 'Seventh Five-Year Plan: An Assessment', 34/1: 18–30.

—— (9 Nov. 1979), 'China's First Environmental Protection Law', 22/45: 24.

—— (1979*a*), '1978 Grain and Coal Output', 22/2: 7–8.

—— (1979*b*), 'Communiqué on Fulfilment of China's 1978 Economic Plan', 22/27: 38.

Bosson, Rex, and Varon, Benison (1977), *The Mining Industry and the Developing Countries* (Oxford University Press, New York), 292 pp.

Brecker, Richard (1990), 'The End of Investment's Wonder Years', *China Business Review* (US–China Business Council, Washington, DC), 17/1 (Jan.–Feb.): 27–9.

British Petroleum Company (1984–93), *BP Statistical Review of World Energy*, various editions (London).

Bureau of Agricultural and Resource Economics (1988), 'Minerals and Metals Markets in China', by Bill Nagle, in *Resource Trends*, Sept. (Australia Department of Primary Industries and Energy, Canberra), 56–64.

Bureau of Foreign Affairs (1978), 'Briefing on China's Agriculture', 18 Sept. (Ministry of Agriculture and Forestry, Peking).

Bureau of Mineral Resources (1988), 'Kimberlites and Diamonds in China', in *BMR Research Newsletter*, 8, Apr. (Australia Department of Primary Industries and Energy, Canberra), 10–11.

Callot, F. (1985), 'Production et consommation mondiales de minerais en 1983', *Annales des Mines* (Paris), 192/7–8–9, 124 pp.

Central Committee of the Communist Party of China (1984), Circular No. 1 (Beijing).

Chen, Dongling, Clements, Kenneth W., Roberts, E. John, and Weber, E. Juerg (1991), 'Forecasting Steel Demand in China', *Resources Policy* (London), 17/3 (Sept.): 196–215.

Chen Nai-ruenn (1967), *Chinese Economic Statistics* (Aldine Publishers, Chicago).

Chen Yunhe (1986), 'Selective Underground Mechanization', *Engineering and Mining Journal* (New York), 187 (May): 31–3.

Cheng, Chu-yuan (1982), *China's Economic Development, Growth and Structural Change* (Westview Press, Boulder, Colo.).

Chesneaux, Jean (1979), *China: The People's Republic, 1949–1976* (Pantheon Books, New York).

Chin, Edmond (1987*a*), 'Far East: China', in *Mining Annual Review 1987* (Mining Journal Ltd., London).

—— (1987*b*), 'The Mineral Industry of China', Preprint from the 1985 *US Bureau of Mines Minerals Yearbook 1985* (US Government Printing Office, Washington, DC), 27 pp.

—— (1986), 'Far East: China', in *Mining Annual Review 1986* (Mining Journal Ltd., London), 351–60.

—— (1985), 'Far East: China', in *Mining Annual Review 1985* (Mining Journal Ltd., London), 381–7.

China Coal Industry Publishing House (1991), *China Coal Industry Yearbook 1990*, Sept. (Beijing), 128 pp.

China Daily [Beijing] (16 June 1993), 'Gold Mines Will Seek Overseas Assistance', 2.

—— (18 Jan. 1993), 'Metal Trader Determined to Join Big Leagues', 1.

—— (17 Jan. 1993), 'SINOPEC Signs $4m. Deal with Honeywell', 5.

—— (11 Jan. 1993), 'Opening up Mines to Overseas Investors', 1.

—— (9 Jan. 1993), 'Rosy Year for Foreign Investors', 2.

—— (28 Aug. 1992), 'Mining Fee Regulations to Improve Resources', 1.

—— (1 July 1992), 'Bidding to Open for Offshore Oil', 2.

—— (28 Feb. 1992), 'Law Sees Dramatic Drop in Mine Wars', 1.

—— (27 Nov. 1989), 'Mineral Management Plan', Business Weekly Section, 4.

—— (17 Apr. 1989), 'Overseas Investments Continue to Increase', Business Weekly Section, 1.

—— (8 Nov. 1988), 'State Bans Licenses for Individual Gold Miners', 1.

—— (17 July 1988), 'Contract Workers', 1.

—— (30 June 1987), 'Australia Invests in Aluminum', by Xu Yuanchao, 2.

—— (29 June 1987), 'Rampant Mining Causes Danger', 4.

—— (15 June 1987), 'Huge Waste of Minerals Pointed Out', 7: 2.

—— (26 Oct. 1985), 'Rural Pollution Warning', 4.

—— (18 Sept. 1985), 'Energy Industry Records Big Growth', 1.

China International Economic Consultants, Inc. (1986), *The China Investment Guide*, 3rd edn. (Longman, London), 921 pp.

China National Nonferrous Metals Industry Corporation [CNNC] (1986), *Brief Introduction to the Enterprises and Institutions of the Nonferrous Metals Industry of China* (Beijing), 327 pp.

China National Offshore Oil Corporation [CNOOC] (1992), *Persisting in the Reform and Open Policy to Develop the Offshore Oil Industry 1982–1992* (Beijing), 188 pp.

China Petrochemical Corporation (1991), *1991 Annual Report* (Beijing), 37 pp.

China Reconstructs [Beijing] (1984), 'Key Projects under Construction', 33 (Apr.): 15–18.

—— (1979), 28/4: 8 and 56.

China Statistical Information and Consultancy Service Centre and the East–West Center (1987), *Input–Output Tables of China, 1981*, compiled by the Centre of Economic Forecasting, State Planning Commission and the Department of Statistics on Balances of National Economy, State Statistical Bureau, Beijing. Distributed by University of Hawaii Press, Honolulu, 107 pp.

Chu Li and Tien Chieh-yun (1974), *Inside a People's Commune* (Foreign Language Press, Beijing).

Clark, Allen L., and Trocki, Linda K. (1991), 'Benefit–Cost Assessment Programs: Costa

Rica Case Study', *Materials and Society*, 15/4: 323–50.

—— Dorian, James P., and Fridley, David G. (1989), 'Problems and Prospects for China's Mineral and Energy Industries', in Charles E. Morrison and Robert F. Dernberger (eds.), *Asia–Pacific Report 1989, Focus: China in the Reform Era* (East–West Center, Honolulu), 75–82.

Clark, Jennifer C. (1988), 'Energy and Minerals Joint-Ventures in China's Development Goals and Foreign Investment', in Ved P. Nanda (ed.), *The Law of Transnational Business Transactions*, i, International Business and Law Series (Clark Boardman, New York).

Cohn, Stanley H. (1970), *Economic Development in the Soviet Union* (Heath Lexington Books, Lexington, Mass.).

Cook, L. H. (1984), 'Quantifying the Effects of Changes in the Minerals Sector: Background Notes', in L. H. Cook and M. G. Porter (eds.), *The Minerals Sector and the Australian Economy* (Allen & Unwin Australia, Sydney), 281–98.

—— and Lees, V. J. (1984), 'Some Quantitative Estimates of the Effects of Changes in the Minerals Sector', in L. H. Cook and M. G. Porter (eds.), *The Minerals Sector and the Australian Economy* (Allen & Unwin Australia, Sydney), 249–79.

Crook, Frederick W. (1975), 'The Commune System in the People's Republic of China, 1963–1974', in United States Congress, Joint Economic Committee Report, *China: A Reassessment of the Economy* (US Government Printing Office, Washington, DC), 109.

Crowson, Philip (1986–92), *Minerals Handbook*, 1986–92 edns. (Stockton Press, New York).

Culliton, Barbara J. (1979), 'China Adopts New Law for Environmental Protection', Science (Washington, DC), 206/4417 (26 Oct.): 429.

David, Michel (1988), *Handbook of Applied Advanced Geostatistical Ore Reserve Estimation* (Elsevier Science Publishers, Amsterdam), 216 pp.

Dernberger, Robert F. (1989), 'China's Economic Reforms', in Charles E. Morrison and Robert F. Dernberger (eds.), *Asia–Pacific Report 1989, Focus: China in the Reform Era* (East–West Center, Honolulu), 53–64.

DeYoung, J. H., Sutphin, D. M., and Cannon, W. F. (1984), *International Strategic Minerals Inventory Summary Report—Manganese*, US Geological Survey Circular 930-A (Washington, DC), 22 pp.

—— —— Werner, A. B. T., and Foose, M. P. (1985), *International Strategic Minerals Inventory Summary Report—Nickel*, US Geological Survey Circular 930-D (Washington, DC), 62 pp.

Dillard, D. (1948), *The Economics of John Maynard Keynes* (Prentice-Hall, Englewood Cliffs, NJ).

Dixon, P. B., Parmenter, B. R., Sutton, J. M., and Vincent, D. P. (1982), *ORANI: A Multisectoral Model of the Australian Economy* (North-Holland Press, Amsterdam).

Dizhi Chubanshe (1985), *Zhong Hua Ren Min Gong He Guo Di Zhi Kuang Chan Bu Nian Jian* (Geology Publishing House, Yearbook of the Ministry of Geology and Mineral Resources China; Beijing), 271–2.

Donnithorne, Audrey (1967), *China's Economic System* (Praeger, New York), 592 pp.

Dorian, James P. (1992), 'Mineral Investment Conditions in China', chapter in *Mineral Investment Conditions in Selected Countries of the Asia–Pacific Region* (United Nations, New York), 35–81.

—— (1988), 'Input–Output Tables of China, 1981', Book review in *Asian and Pacific Population Forum* (East–West Population Institute, Honolulu), 2/1–2: 16–17.

Dorian, James P. (1987), 'The Role of Minerals in the Economic Development of China: 1949–2000', Ph.D. dissertation, University of Hawaii, Honolulu, 253 pp.

—— and Clark, Allen L. (1987), 'China's Energy Resources: Potential Supply, Problems, and Implications', *Energy Policy* (London), 15/1: 73–90.

Du Wenfeng and Wang Qinghua (1992), 'China Expands its Largest Coal Production Base', *Australian Journal of Mining* (Victoria), 7/69 (July).

Economist Intelligence Unit (1990), *China, North Korea: Country Profile, 1990–1991* (Economist Intelligence Unit, London), 55 pp.

The Economist (1992), 'China: The Titan Stirs', 325/7787 (28 Nov.): 1–18.

Emerson, Craig, and Warr, Peter G. (1982), 'Economic Evaluation of Mineral Processing Projects: A Case Study of Copper Smelting in the Philippines', Development Studies Center Occasional Paper 32 (Australian National University, Canberra), 29 pp.

Emerson, John Philip (1982), 'The Labor Force of China: 1957–1980', in United States Congress, Joint Economic Committee Report, 97th Congress, 2nd Session, *China under the Four Modernizations*, Part 1 (US Government Printing Office, Washington, DC), 224–67.

Energy, Mines, and Resources Canada (1988), *The Mineral Industry of China*, ed. A. Werner, May (Resource Evaluation Division, Mineral Policy Sector, Ottawa), 112 pp.

Energy Program (1990), Personal communication, East–West Center, Honolulu.

Far Eastern Economic Review [Hong Kong] (24–31 Dec. 1992), 'Numbers Game', 154/52: 72–3.

—— (12 Nov. 1992), 'China's Oil Shock', 155/45: 52–4.

—— (16 Apr. 1992), 'No Sign of Fatigue', 155/15.

—— (7 Feb. 1991), 'Pit Stop in China', 151/6: 53–4.

Field, Robert Michael (1986), 'China: The Changing Structure of Industry', in United States Congress, Joint Economic Committee Report, 99th Congress, 2nd Session, *China's Economy Looks Toward the Year 2000* (US Government Printing Office, Washington, DC), i. 505–47.

Foreign Broadcast Information Service (25 Aug. 1989), 'Warning on Effects of Ore Theft from Mines', FBIS-CHI-89-164, p. 22.

—— (25 Mar. 1986), 'Text of Mineral Resources Law Released', *China, PRC National Affairs* (US Government Printing Office, Washington, DC), 1/057: K19–K26.

—— (12 Mar. 1986), 'Mineral Resources Law Viewed', *China, PRC National Affairs* (US Government Printing Office, Washington, DC), FBIS-CHI-86-048: K3–K5.

Foreign Language Press (1984), *The Sixth Five-Year Plan of the People's Republic of China for Economic and Social Development (1981–1985)* (Beijing).

Fridley, David G. (1993), Personal communication, East–West Center, Honolulu, Hawaii.

—— (1991*a*), 'China's Changing Oil Industry', Energy Program Petroleum Advisory 77 (5 Sept.), East–West Center, Honolulu, 7 pp.

—— (1991*b*), 'Issues in the Development of Mainland China's Oil Industry', Paper presented at the Seminar on Mainland China's Oil Industry (May), Taipei, Taiwan.

—— (1991*c*), 'China's Energy Outlook', in *China's Energy Dilemmas in the 1990s: The Problems of Reforms, Modernization, and Interdependence*, Joint Economic Committee report (Apr.), United States Congress, Washington, DC, ii. 495–526.

—— (1991*d*), 'China's Energy Outlook', East–West Center Working Paper (Mar.), Honolulu, Hawaii, 53 pp.

—— (1990), Personal communication, Energy Program, East–West Center, Honolulu.

—— (1988), *China: A Survey of Chinese Refining from Toppers to Bottoms*, US Department of Energy Report, Washington, DC, 39 pp.

—— and Fereidun Fesharaki (1987*a*), 'China's Petroleum Industry: International and Domestic Policy Imperatives', China Energy Study (Mar.), East–West Center, Honolulu, 39 pp.

—— —— (1987*b*), *China and the Changing Global Industry* (East–West Center, Honolulu).

General Administration of Customs, PRC (1983–92), *China's Customs Statistics* (Economic Information and Agency, Hong Kong), various issues.

Gao Wanli (1986), 'The Mineral Resources of China', unpublished manuscript, National Geological Archives, Ministry of Geology and Mineral Resources, Beijing.

Geological Publishing House (1992), *Mineral Resources Maps of China, Scale 1:5,000,000*, ed. Song Shuhe, with explanatory text in English (Beijing), 3 maps.

—— (1990), *Mineral Deposits of China*, i (Beijing), 355 pp.

Ghose, A. K. (1990), 'Mining in China—Looking for New Directions', *Asia–Pacific Mining* (Singapore), Apr.–May: 20–4.

Gorst, Isabel (1989), 'Ambitious Offshore Targets', *Petroleum Economist* (London), 61/9 (Sept.): 275–7.

—— (1988), 'More Promising Finds Offshore', *Petroleum Economist*, 60/3 (Mar.): 79–80.

Hayes, David (1986), 'China Takes a Fresh Look at Onshore Natural Gas', *Asian Oil and Gas*, Feb./Mar.: 14.

Hsueh Mu-chiao, Su Hsing, and Lin Tse-li (1960), *The Socialist Transformation of the National Economy in China* (Foreign Language Press, Beijing), 287 pp.

Hu Gengkang, Zhang Tingquan, and Liu Bingwen (1984), *Economy*, China Handbook Series (Foreign Language Press, Beijing).

Huang Jichun (1987), 'The Rising of China's Non-ferrous Metals Industry', unpublished manuscript of speech, China National Nonferrous Metals Industry Corporation, Beijing, 14 pp.

Hufschmidt, Maynard M., James, David E., Meister, Anton D., Bower, Blair T., and Dixon, John A. (1983), *Environment, National Systems, and Development* (Johns Hopkins University Press, Baltimore), 338 pp.

Ikonnikov, A. B. (1975), 'The Mineral Resources of China', microfiche only (Geographical Society of America, Boulder, Colo.).

International Agricultural Development Service (1981), *Agricultural Development Indicators* (New York), 17 pp.

International Iron and Steel Institute (1993), World Steel Consumption Figures, provided by Mr Phil Norris, Statistics Department, Brussels.

Intertrade [Beijing] (1988), 5/5.

Jackson, R. T., Emerson, C. A., and Welsch, R. (1983), *The Impact of the OK Tedi Project*, Report prepared for the Department of Minerals and Energy, Papua New Guinea, Port Moresby, 429 pp.

Jensen, R. C., and West, G. R. (1986), 'Australian Regional Developments No. 1', in Department of Economics, University of Queensland, *Input–Output For Practitioners: Theory and Applications* (Australian Government Publishing Service, Canberra).

—— Mandeville, T. D., and Karunaratne, N. D. (1979), Regional Economic Planning (Croom Helm, London).

Jia Yunzhen (1988*a*), 'The Future Market for China's Coal', in James P. Dorian and David G. Fridley (eds.), *China's Energy and Mineral Industries: Current Perspectives* (Westview

Press, Boulder, Colo.), 59–63.

Jia Yunzhen (1988*b*), 'A Review of China's Coal Export Potential through her Domestic Demand and Supply', Paper presented at The Asia–Pacific Coal Trade Workshop: The Dynamics of Expanding Supply, 29–30 Sept., East–West Center, Honolulu, 14 pp.

—— (1988*c*), 'The Situation of Small-Scale Coal Mining: Present, Future, and its Importance in the Development of National Economy in the People's Republic of China', unpublished manuscript, Ministry of Energy, Beijing.

Johnson, Charles J. (1992), 'Asia's Coal Future to 2010', Paper presented at the Western Coal Transportation Association Conference, 6–9 Apr., Tucson, Arizona, 26 pp.

—— and Pintz, William S. (1985), 'Minerals and the Developing Economies', in William A. Vogely (ed.), *Economics of the Minerals Industries*, 4th edn. (American Institute of Mining, Metallurgical, and Petroleum Engineers, New York), 19–50.

Kang Chao (1970), *Agricultural Production in Communist China* (University of Wisconsin Press, Madison), 139 and 151.

Kuo, Leslie T. C. (1976), *Agriculture in the People's Republic of China* (Praeger, New York).

Lardy, Nicholas R. (1986), 'Overview: Agricultural Reform and the Rural Economy', in United States Congress, Joint Economic Committee Report, 99th Congress, 2nd Session, *China's Economy Looks Toward the Year 2000* (US Government Printing Office, Washington, DC), i. 325–35.

Lavely, William (1989), 'Demographic and Social Change in China', in Charles E. Morrison and Robert F. Dernberger (eds.), *Asia–Pacific Report 1989, Focus: China in the Reform Era* (East–West Center, Honolulu), 65–73.

Lee Kuen (1989), 'An Experimental China CGE Model', Resource Systems Institute Working Paper, East–West Center, Honolulu, 9 pp.

—— Lee, Chung H., and Kim, Won Bae (1990), 'Problems and Profitability of Direct Foreign Investment in China: An Analysis of the Survey Data', *Journal of Northeast Asian Studies* (Institute for Sino-Soviet Studies, Washington, DC), 9/4: 36–52.

Lee, W. W. (1991), 'Recent Coal Situation and Outlook in Korea', Paper prepared for the Coal in Asia–Pacific Workshop sponsored by the East–West Center, held in Bangkok, Thailand.

Levine, Herbert S. (1962), 'Soviet Economic Planning: Input–Output Analysis and Soviet Planning', *American Economic Review* (Nashville), 52: 127–37.

Li Honglin (1986), 'Four Isms of China—Feudalism, Capitalism, Socialism, and Communism', Seminar presented by Member of Fujian Academy of Social Sciences, China, at the East–West Center, Honolulu, 25 June.

Liang Wensen (1982), 'Balanced Development of Industry and Agriculture', in Xu Dixin (ed.), *China's Search for Economic Growth* (New World Press, Beijing), 52–78.

Lin, Cyril Zhiren (1989), 'Open-Ended Economic Reform in China', in Victor Nee and David Stark (eds.), *Remaking the Economic Institutions of Socialism: China and Eastern Europe* (Stanford University Press, Stanford, Calif.), 95–136.

Liu Guoguang and Wang Xiangming (1980), 'A Study of the Speed and Balance of China's New Economic Development', *Social Sciences in China* (Beijing), 4.

Liu Jingtong (1986), 'Energy Access Held Vital to Industrial Development', *China Daily*, 7 July: 4.

Liu Suinian and Wu Qungan (1986), *China's Socialist Development: An Outline History (1949–1984)* (Beijing Review Publishers, Beijing), 700 pp.

Lu Nan and Li Mingzhe (1988), 'Use of Input–Output Techniques for Planning the Price Reform', unpublished manuscript, Price Research Institute, State Price Bureau, Beijing, 19 pp.

Lu Zhen (1988), 'China's Iron Ore Industry', unpublished manuscript, Institute of Industrial Economics, Chinese Academy of Social Sciences, Beijing, Apr. 29 pp.

MacFarquhar, Roderick (1983), *The Origins of the Cultural Revolution*, ii (Columbia University Press, New York), 470 pp.

Mao Zedong (1977), *Selected Works*, v (Foreign Language Press, Beijing).

Metallgesellschaft (1992), *Metallstatistik 1981–1991*, 79th edn. (Frankfurt), 512 pp.

Miernyk, W. H. (1965), *The Elements of Input–Output Analysis* (Random House, New York).

Minerals Policy Program (1993), 'Mineral Investment, Consumption and Production Data Bases', Asia–Pacific Minerals Trade and Investment Project, East–West Center, Honolulu.

Mining Engineering (1992), 'China Will Lift Price Controls on Half of its Coal Output', 44/9 (Sept.).

Mining Journal (4 June 1993), 'Chinese Door Opening?', 320/8227: 402.

—— (5 Feb. 1993), 'Chinese Steel Records', 320/8210: 94.

—— (22 Jan. 1993), 'BHP's Chinese Connection', 320/8208: 59.

—— (15 Jan. 1993), 'Market News', 319/8207: 55.

—— (11 Dec. 1992), 'Chinese Death Toll', 319/8203: 419.

—— (25 Sept. 1992), 'Iron and Steel Projects in China', 319/8192: 218.

—— (28 Aug. 1992), 'Chinese Awakening', 319/8188: 146.

—— (6 Mar. 1992), 'China to Continue Large-Scale Gold Mining', 318/8163: 163.

—— (13 July 1990), 'China's Iron and Steel Industry', 315/8079: 31–3.

—— (24 Mar. 1989), 'China's Gold Increases', 312/8012: 237.

—— (23–30 Dec. 1988), 'Export Ban', 311/8000: 510.

Mining Journal Limited (1982–92), *Mining Annual Review* (London).

Mining Magazine (Nov. 1991), 'China's Coal Problems', 165/5.

—— (May 1991), 'Chinese Exploration', 165/5: 333.

Ministry of Foreign Economic Relations and Trade [MOFERT] (1986–91), *Almanac of China's Foreign Economic Relations and Trade*, various edns. (China Resources Advertising, Hong Kong).

—— (1988), 'China's Status as a Contracting Party: Memorandum on China's Foreign Trade Regime', *Intertrade* (Beijing), 5/1: 26–40.

—— (1987), Collection of Laws and Regulations for the Encouragement of Foreign Investment (in Chinese; Beijing), 23 pp.

Ministry of Geology and Mineral Resources (1991*a*), 'The Development of Geological Surveying, Exploration, and Exploitation of Mineral Resources in China, 1989–1990', unpublished manuscript, Beijing, 8 pp.

—— (1991*b*), 'A Brief Introduction on the Mineral Development Supervision Activities in China', unpublished manuscript, Beijing, 10 pp.

—— (1984), *Contributions from the Various Organizations of the Ministry of Geology and Mineral Resources PRC 1984* (Chinese edn.; Geological Publishing House, Beijing), 88 pp.

—— (1983), *Contributions from the Various Organizations of the Ministry of Geology and Mineral Resources PRC 1983* (Chinese edn.; Geological Publishing House, Beijing), 79 pp.

Ministry of Geology and Mineral Resources (1982), *Contributions from the Various Organizations of the Ministry of Geology and Mineral Resources PRC 1953–1982* (Chinese edn.; Geological Publishing House, Beijing), 71 pp.

Nankani, Gobind (1979), 'Development Problems of Mineral-Exporting Countries', World Bank Staff Working Paper 354, Washington, DC.

Oceana Publications (1982), *Constitutions of the Countries of the World*, iv (Dobbs Ferry, NY).

Odagawa, Keiho (1988), 'The Chinese Steel Industry and China's Economic Modernization', *China Newsletter*, 74 (Japan External Trade Organization (JETRO), Tokyo), 2–10.

Oil and Gas Journal [Tulsa, Okla.] (1988), 'China Chasing Western Downstream Deals', 8 Aug.: 18.

Ottley, G., and Valauskas, E. (1983), 'China's Developing Environmental Law: Policies, Practices, and Legislation', *6 B.C. International and Comparative Law Review*, vol. 81.

Otto, James M. (1990), 'China's Nonenergy Minerals Industry: Taxation and the Distribution of Income', *Materials and Society*, 14/1: 79–101.

Peking Review (10 Mar. 1978), 'Unite and Strive to Build a Modern, Powerful Socialist Country', 21/10: 7–40.

—— (8 Oct. 1976), 'Carry on the Revolutionary Tradition Cultivated by Chairman Mao', 19/41: 17.

Perkins, Dwight H. (1975), 'Constraints Influencing China's Agricultural Performance', in United States Congress, Joint Economic Committee Report, *China: A Reassessment of the Economy* (US Government Printing Office, Washington, DC), 353.

—— (1966), *Market Control and Planning in Communist China* (Harvard University Press, Cambridge, Mass.), 291 pp.

—— *et al.* (1977), *Rural Small-Scale Industry in the People's Republic of China* (University of California Press, Berkeley).

Petroleum Economist (1993), 'China', 60/6 (June): 78.

—— (1991), 'Shooting for the Stars', 58/6 (June): 16–17.

PetroMin [Singapore] (1989), 'The Petroleum Geology of Onshore China—A Concise and Systematic Review', Nov.: 32–5.

Pintz, William S., and Rizer, James P. (1985), 'The Impact and Planning of Large Resource Development Projects', Working Paper, Pacific Islands Development Program, June, East–West Center, Honolulu.

Polenske, Karen R. (1986), 'Chinese Input–Output Research in Western Perspective', Paper presented at the Eighth International Conference on Input–Output Techniques, Sapporo, 28 July–2 Aug. 1986, 47 pp.

Pomfret, Richard (1991), *Investing in China* (Iowa State University, Ames), 152 pp.

—— (1989), *Equity Joint Ventures in Jiangsu Province* (Longman, Hong Kong).

Radetzki, Marian (1982), 'Regional Development Benefits of Mineral Projects', *Resources Policy*, 8: 193–200.

Rawski, Thomas G. (1979), *Economic Growth and Employment in China*, published for the World Bank (Oxford University Press, New York).

Renmin Ribao (1991) [People's Daily; Beijing], 16 Apr.

—— (1990), 2 Apr.

Riskin, Carl (1987), *China's Political Economy: The Quest for Development since 1949* (Oxford University Press, Oxford), 418 pp.

Robinson, S., and Tyson, L. (1985), 'Foreign Trade, Resource Allocation, and Structural Adjustment in Yugoslavia', *Journal of Comparative Economics*, 9: 46–70.

Rose, Adam, and Miernyk, William (1986), *Input–Output Analysis: The First Fifty Years*, Research Paper 8626, Regional Research Institute, West Virginia University, Morgantown, 81 pp.

Ross, Lester, and Mitchell, A. Silk (1987), *Environmental Law and Policy in the People's Republic of China* (Quorum Books, New York), 449 pp.

Saich, Tony (1981), *China: Politics and Government* (St Martin's Press, New York).

Schena, Gianni Davide (1989), 'An Outline of a System for the Quantitative Management of the Mineral Resource Endowment of a Developing Economy', unpublished manuscript, Imperial College of Science and Technology, London, 36 pp.

Shen Yonggan (1988), 'Forecasting China's Steel Products Demand by 2000', unpublished manuscript, Technical Information Centre, Ministry of Metallurgical Industry of China, Beijing, 42 pp.

South China Morning Post (1986), Oct. 19.

State Statistical Bureau (1981–92), *Zhongguo Tongji Nianjian, 1981–1992* (Statistical Yearbook of China, 1981–1992), Chinese and English edns. (Economic Information and Agency, Beijing and Hong Kong).

—— (1982–88), *Zhongguo Nengyuan Tongji Nianjian 1982–1988* (Energy Statistical Yearbook of China, 1982–1988), various Chinese edns. (China Statistical Publishing House, Beijing).

—— (1985), *Zhongguo Gongye de Fazhan, 1949–1984* (Industrial Development in China, 1949–1984) (China Statistical Publishing House, Beijing).

—— (1960), *Ten Great Years: Statistics of the Economic and Cultural Achievements of the People's Republic of China* (Foreign Language Press, Beijing).

Stone, Richard (1970), 'A Comparison of the SNA and the MPS', in *Mathematical Models of the Economy and Other Essays* (Chapman & Hall, London), 201–23.

Su Kaiming (1986), *Modern China: A Topical History* (New World Press, Beijing), 279 pp.

Summary of World Broadcasts (1986), 'Collective and Individually-Run Mines' (Sept. 10), FE/W1406/A/10.

Taylor, Jeffrey R. (1986), 'Labor Force Developments in the People's Republic of China, 1952–83', in United States Congress, Joint Economic Committee Report, *China's Economy Looks Toward the Year 2000* (US Government Printing Office, Washington, DC), 222–62.

Torries, Thomas, Rose, Adam, and Chen, Chia-Yon (1988), 'Mineral Project Selection and Economic Development: The Case of Namibia', *Materials and Society* (New York), 12/3–4: 263–94.

Tregear, T. R. (1970), *An Economic Geography of China* (Elsevier, New York), 208.

Tse, Pui-Kwan (1992), 'China', in Mining Journal Ltd., *Mining Annual Review 1992* (London, June), 84–90.

United Nations (1984–90), *UNCTAD Commodity Yearbooks*, various issues (United Nations Conference on Trade and Development, New York).

—— (1977*a*), *Comparison of the System of National Accounts and the System of Balances of the National Economy*, Part I. *Conceptual Relationships*, Series F, No. 20, E.77.XVII.6. (New York).

—— (1977*b*), *Manual on the Compilation of Commodity Balances* (Economic and Social Commission for Asia and the Pacific, Bangkok).

United Nations (1975), *Standard International Trade Classification Revision 2*, Statistical Papers Series M, 34/Rev. 2 (UN Statistical Office, New York), 102 pp.

United Nations Industrial Development Organization (1991), 'China: Towards Sustainable Industrial Growth', UNIDO Industrial Development Review Series (Blackwell, Oxford), 26 pp.

US Bureau of Mines (1982–93), *Mineral Commodity Summaries*, 1982–93 edns. (Department of the Interior, Washington, DC).

—— (1951–92), *Minerals Yearbook*, 1949–90 edns. (Department of the Interior, Washington, DC).

—— (1948), *Mineral Resources of China*, 2/7 (Jan.), Foreign Minerals Division, Foreign Minerals Survey Series (US Bureau of Mines, Washington, DC), 226 pp.

US–China Business Council (1989–92), *China Business Review*, various issues (China Business Forum, Washington, DC).

—— (1988), *China's Metals and Minerals: A Comprehensive Guide* (Washington, DC, Oct.), 132 pp.

US Department of Commerce (1988), *International Director Investment: Global Trends and the US Role*, International Trade Administration (US Government Printing Office, Nov. Washington, DC).

US International Trade Commission (1985), 'China's Economic Development Strategies and their Effects on US Trade', Investigation 332-168, USITC Publication 1645 (Washington, DC, Feb.).

US National Foreign Assessment Center (1977), *China: Economic Indicators* (Washington, DC), 13 and 20.

Wang, K. P. (1975), *The People's Republic of China: A New Industrial Base with a Strong Mineral Base* (US Bureau of Mines, Washington, DC).

Wang Xionglin (1988), 'Mineral Resources in China: Geological Exploration and Exploitation', in James P. Dorian and David G. Fridley (eds.), *China's Energy and Mineral Industries: Current Perspectives* (Westview Press, Boulder, Colo.), 105–16.

West, G. R. (1986), 'Australian Regional Developments No. 2', in *Input–Output for Practitioners: Computer Software User's Manual*, Department of Economics, University of Queensland (Australian Government Publishing Service, Canberra).

Wiens, Thomas B. (1980), 'Agricultural Statistics in the People's Republic of China', in Alexander Eckstein (ed.), *Quantitative Measures of China's Economic Output* (University of Michigan Press, Ann Arbor).

Woodard, Kim (1987), *Trade and Investment in China's Petroleum Industry: Trends and Projections, 1981–1990* (Petroleum Industry Research Foundation, New York), 260 pp.

World Bank (1990*a*), *Macroeconomic Stability and Industrial Growth under Decentralized Socialism*, World Bank Country Study (Washington, DC), 299 pp.

—— (1990*b*), *China: Between Plan and Market*, World Bank Country Study (Washington, DC), 188 pp.

—— (1989), *China: Economic Memorandum: Macroeconomic Stability and Industrial Growth under Decentralized Socialism*, i–iii (Washington, DC, 12 June).

—— (1985*a*), *China: Long-term Development Issues and Options*, Main report (plus six annexe volumes) (Washington, DC).

—— (1985*b*), China: The Energy Sector, Annexe 3 to World Bank (1985*a*).

World Bureau of Metal Statistics (1989–92), *World Metal Statistics Yearbook* (London).

Wu Kang (1989), 'China's Offshore Oil Prospects: An Economic Analysis', China Energy Program, May, East–West Center, Honolulu, 42 pp.

Xin Dingguo (1988), 'The Present and Long Term Strategy of China', in James P. Dorian and David G. Fridley (eds.), *China's Energy and Mineral Industries: Current Perspectives* (Westview Press, Boulder, Colo.), 43–54.

Xinhua News Agency News Bulletin (1986), 'The Seventh Five-Year Plan of the People's Republic of China for Economic and Social Development (1986–1990)', excerpts, No. 013605, Beijing, 15 Apr.

Xinhua News Service [Beijing] (30 Aug. 1989), 'Chinese Oil, Gas Industries Continue Development', OW3108015589.

—— (18 July 1986), 'China Plans to Modernize 77 Coal Mines', XNA5448A.

—— (15 Apr. 1980), 'Suspension of Construction Projects'.

Xu Zhanqian (1992), 'Baotou: China's Rare Earth Valley', *Australian Journal of Mining*, 7/73 (Oct.), 38.

Xue Muqiao (1982), *Current Economic Problems in China*, trans. K. K. Fung (Westview Press, Boulder, Colo.).

Yeh, K. C. (1967), 'Soviet and Communist Chinese Industrialization Strategies', in Donald W. Treadgold (ed.), *Soviet and Chinese Communism: Similarities and Differences* (University of Washington Press, Seattle).

Zhou Enlai (1959), 'Report on the Adjustment of Major Targets in the National Economic Plan for 1959 and Further Development of the Campaign for Maintaining Production and Practicing Economy', *New China Semi-Monthly* [Peking], 17: 22.

Zhou Guojun (1985), 'The Undiscovered Mineral Resources in China', in *Yearbook of the Ministry of Geology and Mineral Resources* (in Chinese; Beijing).

Zhou Taihe (1984) (ed.), *Contemporary China's Economic System Reform* (in Chinese; Social Science Publishers, Beijing).

Zhu Ang (1988), 'China's Investment Environment, Laws, and Regulations', in James P. Dorian and David G. Fridley (eds.), *China's Energy and Mineral Industries: Current Perspectives* (Westview Press, Boulder, Colo.), 127–36.

Zhu G. (1983), *Zhonghua Rermin Gongheguo Fen Sheng Dituji* (Atlas of the People's Republic of China), Beijing, Ditu Chubanshe, 174 pp.

Zhu Xigang and Chapman, John (1988), 'Rural Reforms in China: Current Challenges and Policy Aims', *Quarterly Review of the Rural Economy*, 10/4 (Dec.): 404–14.

INDEX

Notes: This index is in word-by-word alphabetical order. Page numbers in italics indicate references to tables only and not to text. A full list of abbreviations (used only in subheadings) can be found on page xiv.